D0915218

The George Reserve Deer Herd

Dale R. McCullough

The George Reserve Deer Herd

Population Ecology of a K-Selected Species

THE UNIVERSITY OF MICHIGAN PRESS ANN ARBOR

Published in the United States of America by
The University of Michigan Press and simultaneously
in Rexdale, Canada, by John Wiley & Sons Canada, Limited
Manufactured in the United States of America

Library of Congress Cataloging in Publication Data

McCullough, Dale R 1933–
The George Reserve deer herd.

Bibliography: p.
Includes index.
1. White-tailed deer—Ecology. 2. Mammal populations
—Michigan. 3. Mammals—Ecology. 4. Mammals—Michigan
—Edwin S. George Reserve—Ecology. 5. Edwin S. George
Reserve, Mich. I. Title. II. Title: K-selected
species.
QL737.U55M24 599'.7357 79-14254
ISBN 0-472-08611-1

To my former graduate students
and research assistants, the late

John P. Clark
and
James D. Feist

They were fine men, young and full of promise, and taken much too soon.

Preface

On the morning of December 9, 1933, in the predawn gray, thirty men gathered on the George Reserve and conducted the first drive census of the deer herd. That same year, Aldo Leopold's classic *Game Management* appeared. That six deer could have produced a population of 160 in only six years astounded biologists, and the data became a landmark in the field of wildlife management. I was five days old at the time. All lives are a set of improbable coincidences, and the path which brought me to the George Reserve deer studies was circuitous.

My interest in population and ecosystem matters began during my graduate studies at Oregon State, where I launched into an intensive study of ungulate ecology. I began to explore population manipulation and modeling, two areas that became increasingly more important to my study. Too much of population and ecosystem biology rested upon shaky assumptions, and scattered evidence from many different sources could not be fully integrated. One could manipulate numbers forever without knowing which manipulations were realistic. I began to question simplistic concepts: succession is inevitable, climax is stable, carrying capacity is a given number, energy flows, r is a constant, species have typical life tables, etc. I was impressed with the variability of nature. Gradually it began to dawn on me that one must separate the abstract concept, which exists in the mind as a means of organizing information, from the reality of nature. The search for the general all-encompassing statement might be obscuring as much as enlightening. Indeed, simple concepts could become impediments to further advances because they constrained the field of inquiry. For example, if one was concerned with whether a given community was in a successional or climax stage, the answer yes or no was what one searched for. Yet, in the simplest natural community, hundreds of species and thousands of individuals were struggling to get enough energy and nutrients; to eat and avoid being eaten; to survive and reproduce. Declaring a community as climax or successional would be a rather simple description of the state of affairs.

All of these impinging and interacting thoughts convinced me of the need for a study area that was intermediate between the laboratory experiment in which all variables are controlled and the field study with its myriad of uncontrolled variables. It should be large enough to obtain reasonable sample sizes but small enough to allow things to be comprehended and measured. Furthermore, one must be able to manipulate the system in order to drive the experiment along the desired lines. Finally, one must embark on long-term studies in order to sample the total range of responses of which the system is capable.

After completing my masters degree, I moved on to the University of California at Berkeley to study with A. Starker Leopold. Near the end of my work, The University of Michigan requested an interview. When I arrived in Michigan, the late Justin W. (Doc) Leonard drove me around the George Reserve. The opportunity and the study area I had contemplated lay before my eyes.

The reader will note how closely the overall design of this study resembles the International Biological Program (IBP) Terrestrial Production Projects and I gladly acknowledge this influence. However, my studies were planned and underway several years before the IBP studies were launched, and I hope I have contributed as well as received.

Furthermore, I am aware that others have developed ideas which are similar to those expressed here. A broad, integrative hypothesis is unlikely to be totally new. Particularly, I am aware that the ecosystem thinking of Peter Jordan and Dan Botkin with the Isle Royale vegetation-moose-wolf studies, and that of Glenn Cole and his concept of Yellowstone Park as a naturally regulating ecosystem is progressing along some of the same lines. And Graeme Caughley's (1976) integrative paper on ungulate population theory came to my attention at the time that this manuscript was about ready for the publisher. Although his treatment was based upon theory and mine on the analysis of the empirical data from the George Reserve deer population, the conclusions on population dynamics and management, where they overlap, are in substantial agreement.

Nevertheless, I believe the particular integration outlined here is original. I also recognize that after talking about these ideas with hundreds of colleagues and thousands of students some twelve years before they were published, and with the rapid advance of ecology over the same period, the ideas may seem somewhat less innovative in the twelfth year than they did in the first.

It should be understood that this is a progress report. The experiments are continuing to verify responses, test models, and extend and refine the data set. My thoughts are evolving under the selective pressure of new results, and already I have had to abandon some untenable positions and reverse some cherished opinions. I am steeled to the fact that these adjustments are by no means complete. I recognize that the data set presented here is a relatively rare resource. Modelers are facing the "garbage in, garbage out" problem, since it is much easier to build mathematical models than it is to obtain biological data with which to test them. Lindeman's lake and the Isle Royale vegetation-moose-wolf system have been modeled, remodeled, and submodeled ad infinitum, and I have no doubt that the George Reserve deer population data will follow a similar pattern. I am aware that there is a plethora of mathematical techniques to which the data have not been subjected, and some of these approaches will lead to more, or less, or different conclusions than I have reached.

The audience that this book is intended for is all ecologists. It is based upon the conviction that theoretical ecologists and applied ecologists are not good "biological species." Therefore, I include those applied ecologists specifically known as wildlife managers. Indeed, the last part of the book discusses the results relative to management programs.

This book is not intended to be an encompassing work on the white-tailed deer. Its focus is on the population dynamics of white-tailed deer, the environmental context of those dynamics, and a conceptual framework for population responses of K-selected species. Only the more pertinent literature is cited. A complete review of the biology and the extensive literature on the white-tailed deer is well beyond the scope of this book. I hope the current revision of *The Deer of North America,* edited by Lowell K. Halls, will fill that need.

This book is not intended for the layman, but there is no reason why any intelligent person could not comprehend the material. Clearly the information contained herein is highly relevant to the perennial battles waged over management of large ungulates. The audience that needs to be reached, the old buck hunter, the protectionist, etc., will not be reached by this book. The difficult translation of the complexities shown herein to a format appropriate to public education is a gargantuan task left to the future, and preferably to those better equipped for it than I.

Acknowledgments

Literally thousands of people have helped with this research over the years. To the numerous and mostly anonymous people who have aided in the drive censuses and participated in drive crews for the harvest I give my heartfelt thanks. I am greatly indebted to those early biologists who established the records which have come down to me: E. C. O'Roke passed away long before I had heard of the George Reserve, and others, like Adolph Murie and Fredrick Hammerstrom, I know only from a congenial handshake. Irvin Cantrell and my predecessor, Warren Chase, were major sources of records. The analysis of early deer records by David H. Jenkins has been a particularly useful source to me. The first reserve caretaker, Lawrence Camburn, maintained many of the earliest records, and they were continued by subsequent caretakers Joe Griffiths and Dick Wiltse. Dick Wiltse, the current caretaker, and his crew have been especially helpful, and they have assumed the considerable job of harvesting deer. My thanks to the many maintenance crew people who have worked so ably in the dressing station, and the shooters who kept them busy.

For my tenure of research I am thankful to successive George Reserve directors Theodore Hubbell, Nelson Hairston, and Donald Tinkle for their encouragement and support, and especially to Associate Director Francis Evans for his continuing interest, cooperation, and patience.

I gladly acknowledge my indebtedness to Fred E. Smith and John A. Kadlec (director and coordinator of the IBP Analysis of Ecosystems Program) for the many fruitful discussions and valuable ideas they contributed.

Deep felt appreciation goes to my colleague Archibald B. Cowan for the yeoman service he has performed in managing the deer herd, processing deer, and collecting data; but more, I appreciate his hearty encouragement, ready smile, and unflagging spirits.

The Michigan Department of Natural Resources contributed to this project in many ways—from patrolling the boundary fence to participating in the drive censuses to facilitating the obtaining of the state of Michigan Game Breeders License. The department's assistance in the deer research and the importance of the results to its management programs have resulted in a symbiotic relationship. Of particular help have been David Jenkins, Ralph Blouch, Larry Ryel, and Lee Queal.

To my graduate students and/or research assistants, I owe a great deal for their companionship and sharing the work: Dennis King, Dennis Bromley, Fred Samson, Bruce Coblentz, Michael Collopy, David Hirth, David Kitchen, John P. Clark, Jim Feist, Edwin Chinn, Yvette McCullough, John Bissonette, Steve Newhouse, John Wehausen, Peter Flanagan, and Terry Bowyer.

For reading the manuscript and offering many suggestions for its improvement, I wish to thank John Bissonette, Ralph Blouch, Bruce Coblentz, Nancy Green, David Jenkins, David Hirth, John Kadlec, David Kitchen, James Koplin, L. L. Master, Yvette McCullough, Philip Myers, John Ozoga, Larry Ryel, and Louis Verme.

The National Science Foundation has provided continuing financial support under grants GB–6171, GB–12958, GB–28822X, GB–41139, and GB–41139/7300787.

Contents

Introduction

The specific data set which forms the basis of this book is from a long-term study of the white-tailed deer (*Odocoileus virginianus borealis* Miller) population on the Edwin S. George Reserve in southeastern Michigan. However, the principles derived are general and should apply to most large, long-lived mammals—in modern ecological parlance, K-selected species.

Two fundamentally different strategies of survival have been recognized in organisms, and they have been termed K-selected and *r*-selected (MacArthur and Wilson, 1967; Pianka, 1970; Stubbs, 1977). K-selected species are those which have been adapted to live in relatively stable habitats at population densities at or near carrying capacity. This strategy favors evolution of strong competitive ability, which is usually expressed in large body size, long life span, low reproductive rate, and high parental care. By contrast, *r*-selected species are those selected for maximum rate of increase. Typically they are small, have short life expectancy, produce massive numbers of offsprings, and give little parental care (e.g., most insects, microbes, etc.). They usually live in temporary or uncertain environments and are adapted to rapid population growth when conditions are favorable. They tend to show boom or bust population responses and seldom achieve equilibrium at carrying capacity.

Most population theory rests upon the responses of *r*-selected organisms. Unsurprisingly, there are problems in applying this theory to K-selected species (Miller, 1976). The reasons *r*-selected organisms predominate in population studies is obvious. Meal worms are much more easily crowded into an experimental bottle than are deer. Furthermore, the number of generations one might study as a Ph.D. candidate with *r*-selected organisms might require a life time (perhaps several life times) with most K-selected organisms. Granting agencies might balk at the food budget for a colony of elephants, yet this kind of animal plays an important food and recreational role in human societies. Certainly of more than passing interest is the fact that man, himself, is numbered among the K-selected species of the earth.

While the reasons that population ecology is biased toward *r*-selected species are easily understood, it is less obvious why these results have limited applicability to K-selected species (Miller, 1976). The major difficulty appears to arise from the implicit assumption that all of the individuals in the population are qualitatively equal. Even the casual student of genetics will recognize that this assumption is not true, even for *r*-selected organisms. Most species show an appreciable amount of genetic variation, and even genetically identical individuals show phenotypic differences. Nevertheless, with *r*-selected species, this assumption is reasonably well approximated, and ignoring the qualitative variable does not create great havoc with the model. Ignoring qualitative differences in K-selected species can result in problems. Typically, these populations exist in relatively low numbers in a rather close relationship to resources. Competitive relations are intense, and density dependence touches all aspects of the life cycle. Qualitative differences among individuals in the population are enormous. They include differences due to age, sex, and health, and these factors influence survivorship and reproduction. These animals typically possess individual identities in highly complex social structures. Social standing, as expressed through dominance hierarchies, systems of territoriality, etc., mediates competitive interactions and further influences survival and reproduction. Extended parental care influences not only size and condition of offspring, but frequently social status as well. Social alliances based upon kinship or mutual interest crop up in unexpected ways. These factors have important population consequences, and they moderate the numerical responses one would expect if all individuals in the population were qualitatively equal. No population parameter can safely be assumed to be a constant.

It is the effects of these qualitative differences with which this book is mainly concerned. Beginning with the specific example of the George Reserve deer herd, for which the early results of a long-term set of experiments are reported, it moves to the generalizations that can be derived and their applications to most K-selected organisms. Finally, the implications for management of K-selected species in the real world setting are considered.

A secondary theme concerns the ecosystem framework in

which these events occur. This is particularly important for K-selected species which tend to have populations existing near or at carrying capacity. Therefore, it is necessary to know about the resource base in order to interpret population responses. In effect, nested and interrelated population response and ecosystem function hypotheses are being tested in this series of experiments. The reason for highlighting ecosystem considerations is to give context to the deer population dynamics. Detailed results on weight and growth, food habits, consumption and assimilation, flux in energy and nutrients, production, standing crop, and chemical composition of vegetation and the like will be presented elsewhere.

Description of the George Reserve

The George Reserve is a 464-ha (1,146-a) area located in Livingston County, approximately 7.7 km (3 mi) west of Pinckney, Michigan. In the 1920s a Detroit industrialist, Colonel Edwin S. George, purchased twelve adjoining farms to form an estate. In 1930 he gave the property to the University of Michigan with the provision that he be allowed to retain the house in the center of the reserve and the surrounding 16 ha (40 a) for his own use. Following his death in 1940, his heirs relinquished this right and the university assumed complete control.

Colonel George stipulated in his gift that the area should be allowed to follow its natural course without interference by man, thereby serving as a natural laboratory for scientific study. In keeping with this charge, the reserve has been administered by the university's Museum of Zoology as a natural area for scientific research, and artificial controls have been kept to a minimum. For a fuller treatment of the history of the area, see Cantrell (1943).

The climate of the reserve, based on weather records at the center of the area, is typical of southern Michigan—relatively hot, humid summers and moderately cold winters. The hottest months are July and August, with mean maximum temperatures of 26.7°C and mean minimums of 14°C; extreme highs are around 32.2°C. The coldest month is January, with mean maximum and minimum temperatures at −1.4°C and −10.6°C. Variation in temperature is greater in winter than in summer, and extremes of −25°C are relatively common, although they usually last only a few days.

Precipitation does not vary greatly from year to year. It averages 76 cm per year, 70 percent of which occurs between April and

October as rainfall. The greatest rainfall occurs during June and July, but precipitation occurs throughout the year. Snowfall is highly variable from winter to winter: the mean annual snowfall for twenty-one years (1950–71) was 106.2 cm, but it ranged from 13 to 213 cm. Similarly, the number of days with snowcover averaged 76.9 per winter, but ranged from 0 to 116 days.

The landscape shows a relatively rough topography for southern Michigan because of the strong influence of glaciation. Steeply sided ridges enclose numerous wetlands that include kettleholes, leatherleaf bogs, fresh water marshes, and tamarack swamps. Steeper slopes are wooded, while gentler slopes were originally cleared for pasture and croplands. However, the soils are sandy and gravelly, and the low fertility of old fields resulted in their reversion to pasturage before Colonel George acquired the land. Thus, while the soils are generally poor, the reserve contains a remarkable diversity of topography and vegetation that is broken into small units and greatly interspersed (figs. I.1 and I.2).

The most recent and complete mapping of vegetation of the reserve was done by Roller (1974). He described eight major vegetation types: hardwood forest, pine forest, grassland, tamarack swamp, bog, marsh, swamp margin, and forest ecotone. Following are the various cover types (see fig. I.2) and the land area percentages they comprise. Open water accounts for 1.07 percent of the area. Hardwood forests constitute 46.93 percent of the reserve, and the primary tree species are black oak (*Quercus velutina*), red oak (*Q. rubra*), white oak (*Q. alba*), shagbark hickory (*Carya ovata*), pignut hickory (*C. glabra*), black cherry (*Prunus serotina*), red maple (*Acer rubrum*), sugar maple (*A. saccharinum*), big tooth aspen (*Populus grandidentata*), and sassafras (*Sassafras albidum*). The shrub layer consists of a regeneration of the tree species along with huckleberry (*Gaylussacia baccata*), witch hazel (*Hamamelis virginiana*), rose (*Rosa* spp.), June berry (*Amehanchier* spp.), and many others. The herbaceous layer is dominated by sedge (*Carex pensylvania*), with hepatica (*Hepatica triloba*) as a common forb.

Ecotones are resulting from the gradual encroachment of forest regeneration along the forest-openland boundaries, and this type accounts for 1.82 percent of the area. The three species of oaks predominate, but hickories and black cherry also occur.

Pine forests consist of one red pine (*Pinus resinosa*) plantation which makes up 0.11 percent of the area.

Grasslands, which make up 25.61 percent of the area, occur on the level areas and gentler slopes that were cleared for fields and pastures. Major grasses are Kentucky bluegrass (*Poa pratensis*), Canada bluegrass (*P. compressa*), big bluestem (*Andropogon scoparius*), little bluestem (*A. Gerardi*), panic grass (*Panicum* spp.), pussytoes (*Antennaria neglecta; A. plantiginifolia*), yarrow (*Achillea millefolium*), and mint (*Monarda fistulosa*), plus many other species in lesser amounts. In addition, hawthorns (*Crataegus crusgalli*), red cedar (*Juniperus virginiana*), low juniper (*J. communis*), and apple trees (*Malus pumila*) occur as scattered individuals in the openlands, as do clones of smooth sumac (*Rhus glabra*), gray dogwood (*Cornus racemosa*), and trembling aspen (*Populus tremuloides*).

Tamarack swamps comprise 12.88 percent of the area and the dominant trees are tamarack (*Larix laricina*) and poison sumac (*Rhus vernix*). Shrubs include red-osier dogwood (*C. stolonifera*), silky dogwood (*C. amomum*), gray dogwood, swamp birch (*Betula pumila*), leatherleaf (*Chamaedaphne calyculata*), and chokeberry (*Pyrus arbutifolia*). A very sparse understory of sedges and forbs occurs in the swamps.

Swamp margins constitute 2.91 percent of the area. Major species include scarlet oak (*Q. coccinea*), jackoak (*Q. ellipsoidalis*), red maple, black cherry, sassafras, yellow birch (*Betula lutea*), ash (*Fraxinus* spp.), cottonwood (*Populus deltoides*), hickory, black oak, and aspen.

Marshes are dominated by cattails (*Typha latifolia, T. augustifolia*) and sedges (*Carex* spp.), and are lined along their margins by red-osier, silky and gray dogwood, and willows (*Salix* spp.). They comprise 7.62 percent of the area.

Bogs constitute 1.05 percent of the area and are predominately leatherleaf, with sphagnum moss, blue berry (*Vaccinium* spp.), and sedges intermixed. For a more complete recent description and discussion of the changes in the vegetation over the years see Roller (1974).

History of the Deer Population

Deer had been extirpated from the area prior to the time that Colonel George purchased the land. Indeed, there is question if deer occurred at all in southern Michigan at that time. Some observers are of the opinion that escapees from the George Reserve served as one source for development of the present deer herd in southern Michigan (McNeil, 1962).

FIG. I.1. Oblique aerial view of the central part of the George Reserve taken from the south. Note the boundary fence in foreground, rough terrain, and interspersion of vegetation types. (Photo by James W. Wheeler, winter, 1949.)

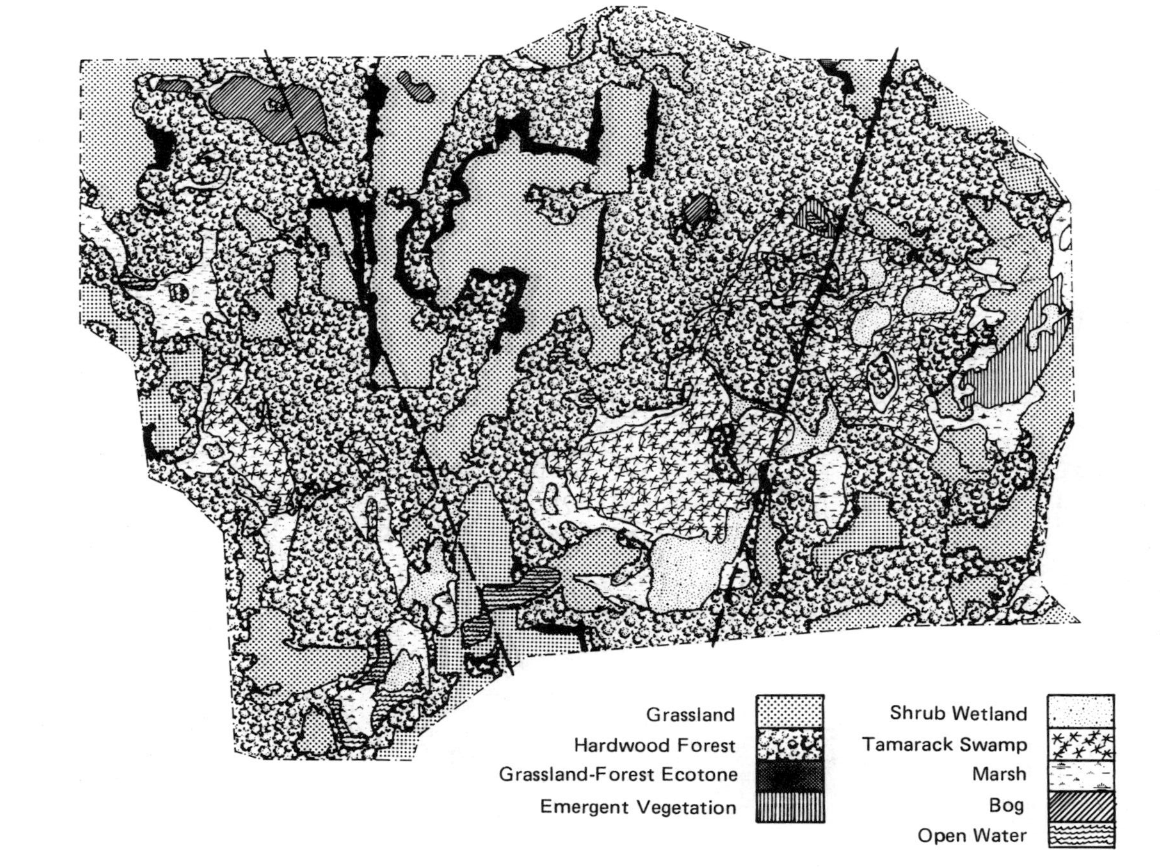

FIG. I.2. Vegetation map of the George Reserve in 1971 (after Roller, 1974). The vertical dashed lines indicate the area covered by the photo in figure I.1.

The reserve deer herd had its genesis in 1928. A 7.5-ft (2.9 m) fence around the reserve was completed in 1927, and the following year Colonel George introduced six deer—two males and four females—that he purchased from the Cleveland-Cliffs Company on Grand Island in Lake Superior. The females were thought to be pregnant.

The population boomed in the following years, providing a classic example of the breeding potential of a white-tailed population in a favorable habitat. In 1933, the first annual deer drive census was held. To the astonishment of the biologists, 160 deer were counted (Hickie, 1937; O'Roke and Hammerstrom, 1948). The number far exceeded the maximum thought possible and this example became a keystone in deer management theory for the following two decades.

Since the reserve was surrounded by a game fence and because the original deer were privately purchased, the state of Michigan issued a Game Breeders License to the university. This legal status, which in effect makes the herd the private property of the university, has continued to the present day and is one of the fundamental keys to the successful conduct of deer population studies. As a private herd, the number and composition of animals to be removed could be decided by the researchers, without constraint of state game regulations. Similarly, the means of removal was at the option of the investigators. The importance of this latitude in design and conduct of a research program without constraint of state hunting regulations or collecting permits cannot be overemphasized. The advantages should become apparent upon the reading of this report.

The increased deer population in the mid-1930s began to have a dramatic and highly visible impact upon the vegetation of the reserve. A highline of browsing was observable on several species and was particularly apparent on red cedar and hawthorn. It became obvious to biologists that letting the situation run its course would result in devastation of the vegetation and destruction of other research opportunities. It was imperative that the deer population be artificially controlled by man, even though such action ran counter to the basic philosophy of noninterference in the natural processes of the area. It was recognized that part of the problem was the lack of large predators in the area, and the role of the predator had to be played by man.

Early management of the deer herd is presented in detail by O'Roke and Hammerstrom (1948), Chase and Jenkins (1962), and Jenkins (1964). Only a brief synopsis is presented here.

Drive censuses in the early and mid-1930s showed a continuing large population, even though a small number of animals were shot by the resident caretaker of the reserve. At this stage, it was not appreciated how large a kill was necessary to reduce the population. Between 1937–38 and 1941–42, kills were substantial enough to result in a decrease in the population. Thereafter, the population showed fluctuations between periodic lows and moderate levels.

In 1945 Dr. Warren Chase joined the faculty of the Department of Wildlife Management at the University of Michigan, and shortly thereafter he assumed control of the deer management program. He introduced several important changes, including a scheme of management that went beyond simple herd control and harvest by a driving crew moving deer past positioned shooters. Most importantly, in 1952 he began collecting the lower jaws of all deer killed or found dead, facilitating age determination and estimation of herd size by the reconstruction method (Jenkins, 1964). This jaw collection, which has continued to the present, forms the basic core of data upon which this book rests (see chap. 2).

During the late 1950s, considerable resentment was building among local hunters against the "private hunting club" aspect of the reserve deer management program. They perceived a select group of university professors obtaining sport recreational hunting at public expense. Some deer were observed jumping out of the reserve under the pressure of the annual drive census. Local hunters believed that deer from outside jumped into the reserve under hunting pressure during the regular fifteen-day hunting season in November. Since the drive census was held in early December, and the hunting in the reserve commenced thereafter, local residents believed that shooters on the reserve were harvesting not only the reserve population, but also deer from the outside.

In 1962 the reserve deer harvest was brought to a halt. A local resident threatened suit against the university and the state Department of Conservation (now the Department of Natural Resources) for noncompliance of the fence to regulation height. The Game Breeders License calls for an 8-ft (2.44-m) fence, and the reserve fence was only about 7.5 ft. In the summer of 1963, the fence was raised to its present 11.5-ft (3.51-m) height, which has effectively stopped ingress and egress of deer, and thereby literally plugged one more hole in the data.

Because of the provisions of the Game Breeders License, it has been possible to sell the meat and hides of harvested deer. Sale of the meat has been to private individuals by the dressed half-carcass (current prices are tied to the wholesale price of beef by the half). Hides are sold by the lot of a year's harvest. Several thousand dollars are realized annually from these sales. The income goes into the George Reserve Fund and is used to support all kinds of research on the area.

CHAPTER 1

Rationale

Problems with Current Ecosystem Concepts

Many ecosystem researchers tend to view either implicitly or explicitly the whole ecosystem as the unit upon which natural selection is operating. The Clementsian concept of the community as a discrete entity (Clements, 1916; see review by McIntosh, 1963) and the group selection hypothesis resurrected by Wynne-Edwards (1962) are ideas that die hard. It would be simpler to understand how nature operates if these concepts were true—it would be a luxury not to have to cope with the myriad of variations over time and space of individual organisms of incalculable numbers.

However, the works of Hamilton (1964), Williams (1966), Weins (1966), Brown (1969), Wilson (1975), Barash (1977), and a host of other students of natural selection have demonstrated the unlikelihood of selection at higher levels of organization. In spite of the elaboration of shared genes in kinship groups (Hamilton, 1964), and even exceptions in possible cases of reciprocal altruism in closely knit social groups (Trivers, 1971), the evidence is overwhelming that the basic unit of selection is the individual. Since this is the unit by which genes are packaged, perhaps the only thing that is surprising is the amount of controversy, time, and energy it took to return to the least common denominator.

Recognition of the individual as the unit of selection should have a pronounced cautionary effect upon scientists working at higher levels of organization. Regulation is the outcome of the sums of the individual organismal responses to the environment, including each other. Moen (1973) has given a thorough review of variation between deer due to age, sex, condition, etc., and the importance of these differences in relation to the environment. It must be emphasized that the natural ecosystem is a totally abstract concept without a real (although often perceived as such) organic counterpart. When we describe ecosystem structure and function, we are either describing the state of the system at one moment in time and space or giving a statistical description of stochastic processes which characterize one or a few dimensions of an assemblage of organisms and the abiotic environment. It may be possible to describe patterns, limits, and tendencies of ecosystem development and function, and these are useful results (F. E. Smith, 1975), but one must be constantly on guard against reasoning that is justified by "it's good for the system" or "the system regulates." This is not merely a semantic point. It has to do with confusing causes with outcomes. Knowledge of the ecosystem is not likely to be served by beginning from an erroneous premise.

It is understandable that biological scientists should look to physical systems for models. One cannot help but be impressed that the knowledge of pressure and resistance has resulted in the elaborate water and electrical system that anastomose modern human civilizations. These systems behave with such predictability that engineers can design elaborate schemes and count on them working as planned (give or take a blackout here and there). Surely such predictability is the dream of every applied ecologist in managing natural environments. Yet, if one looks at the practices of these fields, one is overwhelmed by the inadequacies of current models to predict outcomes.

Part of the failure is due to the inadequacy of physical models as analogies for biological models. While most physical models assume various rates in the system to be constant, in biological systems they are variables. For example, in the George Reserve deer population there is no parameter that does not vary, depending upon how the population is managed, be it age of maturity, growth rates, reproduction rates by age class, survival rates, or even timing of the breeding season. Similar variations are apparent in the vegetation. Thus, an apparent constancy in an observed rate is not inherent, but rather is the outcome of multiple constraints. Release or increase the force of the constraint, and the apparent constant expresses itself as a variable.

The second problem concerns energy flux through the system; because it is unidirectional, and perhaps because of a poor choice of terminology, an erroneous impression has developed. Ecologists speak so glibly about energy flow that it is necessary to emphasize that energy does not "flow" in natural ecosystems. It is located, captured or cropped, masticated, and digested by organisms at the expense of a considerable performance of work. Far from flowing, it is moved forcibly (and sometimes even screamingly) from one trophic level to the next. Most ecosystem models are analogous to water flowing down a stairway. Recognition that the water actually is being pumped from compartment to compartment gives a considerably different view of the operation of the system.

The third problem is that most models fail to recognize that interactions between trophic levels occur in both directions. Usually, only the unidirectional energy flow is considered. I will use the term *influences* to cover these nontrophic effects. Influences include such things as dieback of clipped shoots, trampling, incomplete consumption of carcasses by predators, etc. Influences are far more important as ecosystem variables than is commonly appreciated. Consider a disease: the impact on the host is out of all proportion to the amount of energy consumed by the disease organism; or consider forest pest insect outbreaks which kill numerous trees, or beavers building dams on streams. Most of the biomass of vegetation cut down by the beaver is not consumed, but whole new communities of organisms develop around the impoundment created. Only a small fraction of the energy contained in the forest was consumed by these herbivores, and the flow of energy between trophic levels is totally out of proportion to the devastation suffered by the trees. Impact on the system must be measured by other criteria.

Similar influences occur in the transfer of energy from prey to predators. Assume a cottontail rabbit is chased by a fox and makes it to the bramble tangle a bound before the fox. No energy has transferred between trophic levels, but energy was expended by both rabbit and fox. Yet, even measurement of the expended energy fails to assess the situation. The cost to rabbits of having foxes in the world is eternal vigilance. This costs energy directly as well as indirectly by using time that otherwise could have been spent feeding. It also means that rabbits must have escape coverts which are foxproof. They dare not venture too far from their haven, a distance conditioned by how far foxes can be seen approaching. Excellent rabbit food may exist beyond this safe distance (because the rabbit hasn't been there), but the rabbit dares not venture so far. Thus, transfer of energy from vegetation to rabbit is influenced by foxes. An epidemic of rabies in foxes could strongly influence the rate of transfer. Similarly, fire or a farmer's plow might influence the growth of bramble tangles and alter the balance between fox and rabbit. Not even the most precise determination of energy relationships would characterize this interaction. Still, these are the conditions under which the struggle between vegetation, rabbit, and fox is waged. These are but a few examples, and I submit that virtually every organism in the ecosystem presents similar complexities that are not reflected by the energy "flow" approach.

A fourth difficutly is that most ecosystem models presume that energy per se is the regulating factor. Such an assumption ignores other important variables. Growth of a given population of herbivores may be stopped by insufficient quantities of a trace mineral even though energy abounds. Inclement weather may devastate an insect population despite excellent food conditions. These circumstances can be described by the energetics of the system, but the energetics will not explain why the particular outcome occurred.

An additional difficulty with most ecosystem models is that they assume steady states, such as are found in climax situations. However, climax terrestrial communities hardly exist anymore. Most terrestrial ecosystems are kept in a continuous state of fluctuation by the activities of man. Man's influence is, in fact, the major motivation for the development of predictive models. Furthermore, there is accumulating evidence that disturbance, both natural and caused by aboriginal man, was far more common in pre-Columbian times than is commonly supposed. Description and/or modeling of an ecosystem in a given state is useful, but it often tells little about the behavior of the system in an altered state.

An Ecosystem Hypothesis

The ecosystem function hypothesis being tested by this work has been presented briefly elsewhere (McCullough, 1970). It stems partly from the observations of Margalef (1963) and Odum (1969) that as an ecosystem develops through successional stages, the ratio of production to standing crop biomass (P/B ratio) declines. Thus, early successional stages have a low standing crop biomass but a high rate of production relative to the standing crop biomass. As the climax stage is approached, the ecosystem is characterized by a high standing crop biomass and a relatively low rate of production. This observation seems to be verified by a number of studies on a variety of ecosystems.

The concept of the ecosystem hypothesis developed here can be illustrated by the simplest case of plant, herbivore, and predator.

The conceptual model is presented most easily in teleological terms, although the reader should not infer intent other than convenience of presentation. Standing crop biomass within a compartment tends to accumulate to the subsistence level. At low biomass, most energy is used for growth (net production), while at high biomass, most incoming energy is used to maintain existing biomass (zero net production = subsistence). This is because the older individuals can maintain the status quo competitively (Kabat et al., 1953; Ozoga, 1972). Therefore, the inherent tendency of an ecosystem compartment will be to accumulate a high biomass at a maintenance level rather than a low biomass with a rapid growth and turnover. In animals showing deterministic growth (i.e., individuals tend to reach an upper plateau of body size), the population will consist primarily of adults because they are stronger competitors for food than are juveniles, and failure to accomplish adequate growth results in death of the young animals. In organisms showing continuous growth (i.e., individuals can continue to increase in size throughout life), young individuals may survive in a stunted condition. In either case, population growth ceases and standing crop biomass exists on a maintenance basis.

In effect, each compartment tries to maximize energy storage by maximizing its input rate and minimizing its output rate. This is accomplished by the selective pressure for more efficient mechanisms of locating, capturing, and assimilating food and the avoidance or inhibition of consumption by enemies. The latter mechanisms include such things as distasteful or poisonous substances and/or spines of plants, and predator avoidance in animals.

It is apparent from the diagram that a compartment is affected not only by the energy captured from previous compartments and lost to subsequent compartments, but also by a set of other influences. The total impact of a given compartment, including both consumption and influence, is to decrease the standing crop biomass, thereby increasing the production of the previous compartment. In effect, the consumer compartment builds its own biomass while inducing a higher rate of turnover in the food compartment, a self-reinforcing process within limits. Limits are set by the ability of the food compartment to respond to increased impact. If the compartment can withstand the pressure, a steady state of equilibrium will be achieved. If it cannot, time lags will occur in the reaching of new balances.

Inherent characteristics of the interaction between the consumer and the consumed usually set limitations (referred to as feedback mechanisms) on the interaction. Pimentel (1968) has reviewed this topic in relation to *r*-selected organisms and emphasized genetic changes through selection. Thus, while the genetic aspects of feedback mechanisms are frequently mentioned, there are a corresponding set of phenotypic mechanisms that are less well known but which may be of even greater importance in K-selected species. For example, of the total standing crop of green plants, only a part is of high enough quality to be food for herbivores. In general, quality is highest in meristematic tissues and lowest in structural tissues. Even in ruminant animals that use microorganisms to digest cellulose there are limitations set by the high energy required to masticate structural components and the slow rate at which digestion occurs. Also, protein and other nutrients must be present in large enough quantities to sustain the microorganisms and, subsequently, the host.

Quality relative to food availability is a variable. Since the highest quality food is selected, a small population of herbivores consumes a small amount of high quality forage to sustain high rates of production. A dense population of herbivores consumes a greater weight of food with a lower average quality, until the nutrient levels no longer sustain growth but only maintain the existing biomass. Furthermore, seasonal fluctuations in food quality reinforce this limitation, particularly affecting species that do not have dormant stages.

Food quality varies not only with the anatomical part of the plant, but between individuals (i.e., among clones, sprouts versus seedings, shaded versus unshaded, etc.) and between species of plants. For each species of herbivore, some plants are highly palatable while others are never consumed. At low densities, most food is obtained from favored plants. However, at high densities less-favored plants are consumed as more palatable ones are exhausted.

Similar constraints pertain to the interaction between prey and predators. Killing by predators reduces the prey population and increases subsequent searching time to locate prey. With lowered density, individuals of prey population show greater size (less subject to attack if the predator is of similar size), better health (which results in greater alertness and strength for escape), and less susceptibility to predisposing factors to predation, such as malnutrition, parasitism, disease, etc.

Control of the size of the compartment can come from either the energy-nutrient pathway, the influence-consumption pathway, or from the physical side of the system, most commonly weather. In most cases the outcome is the result of varying impacts of all three. While one factor may predominate under a given set of circumstances, at a different state of the system the predominant impact may shift to another. Similarly, the nature of the organism will greatly modify the relative impact of each of the pathways. Inimical weather that results in great mortality in an insect population may have little measurable effect upon a mammal population.

Because the control pathways tend to operate in a compensatory fashion, a great impact by one usually results in a minor impact by the other two. If weather severely reduces a population, neither predators nor food limitations will be very important. If favorable weather prevails, then predators or food or both factors will limit the population. Food limitation is the ultimate control if the others fail.

The basic value of this conceptual model is that it interlinks the system by demonstrating how an impact at one point affects the rest of the system. It assumes that the ratio of production to standing crop biomass that characterizes the total system over succession change (Margalef, 1963; Odum 1969) is true of each compartment and population, as well. In addition, the relationship between levels has qualitative as well as quantitative elements, and both are important in regulation of energy between trophic levels.

A further advantage of this model is that it suggests that many states of the system can be steady states. Rather than a simple equilibrium, a range of equilibria are possible within the limits of the compensatory mechanisms of the system. This encompasses a wide range of states—from energy passing quickly through a series of rapidly cycling levels to one or more levels showing significant accumulations of biomass while others may cycle rapidly. In terms of yield of useful products for human use, the rapidly cycling systems usually are more productive, and the typical strategy of management is to maintain rapid cycling as typified by successional states.

Application of the Hypothesis to White-Tailed Deer

While previous examples used the entire trophic level for each compartment, the same relationships would hold for a given population within a compartment. A subset of particular relevance can be selected in order to obtain a more convenient set of relationships for study. In this study the deer population is the focal point, and those relationships of particular importance to the deer are traced through the system. The interlinking of white-tailed deer, in general, with the conceptual ecosystem model will illustrate the workings of the model and describe the George Reserve studies.

Assume a pristine situation in which a climax equilibrium has been reached by the ecosystem. Deer do best under subclimax conditions, when food is most readily available (Leopold, 1950); thus a climax vegetation is not the best deer range. But assume that deer are able to survive readily on the "edges" created by river courses, blow down, and similar natural and localized factors. Under these conditions, vegetation, deer, and their predators would come to a dynamic, rather than strict, equilibrium in which fluctuations are strongly dampened (fig. 1.1*A*). Most of the plant biomass would be located in forest trees. Only a small portion of the total plant biomass would be within deer reach and, of that, only a small portion would be of suitable quality for deer food. Deer would select this subset of the vegetation and have a rela-

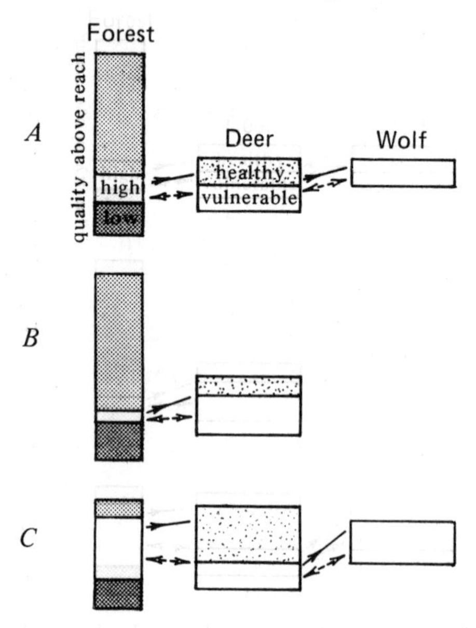

FIG. 1.1. Hypothesized relationship of trophic levels under three system states. *A*. Climax forest. *B*. Climax forest with predators artificially removed. *C*. Subclimax forest. Scales are relative rather than absolute.

tively great impact upon it. Thus, while most of the climax vegetation would be characterized by high standing crop biomass and low production, the subset would show low biomass and high production. This rapid cycling would be reinforced by heavy deer browsing, producing hedging of bushes and a "lawn" of herbaceous vegetation. Particularly palatable species or species vulnerable to grazing would be eliminated or restricted to refugia (the middle of tangles or thorny bushes, etc.). Staple plants in the deer diet would be those species that can withstand heavy defoliation. Typically, they are relatively abundant and moderate in food quality.

In this climax state, the deer population would be small and characterized by an age structure skewed to the older age classes. Because of heavy intraspecific competition for the limited favorable forage, survivorship of fawns would be poor and recruitment low. Therefore, the deer herd would exist in a state where most of the resources were going to maintain a population of adults at the expense of reproduction and the population would have a low P/B ratio.

Still, the deer herd would not be strictly at the subsistence level because of the impact of predators. The wolf (*Canis lupus lycaon*) was the predominant predator in the Great Lakes states in pristine times. Studies by Mech and Frenzel (1971) in Minnesota and Kolenosky (1972) and Pimlott et al. (1969) in Ontario have clearly shown the selective nature of wolf predation. Wolf kills are predominantly the young of the year, the old, and the infirm (see chap. 12). Therefore, the interface between deer and predator is strongly conditioned by quality. Large, healthy deer are not very vulnerable; small, weak deer are highly vulnerable. Similar selectivity by wolves on other prey species have been reported by Crisler (1956), Murie (1944), and Mech (1966). Indeed, purely on theoretical grounds, one would have to hypothesize that a quality factor had to be present to account for the coexistence of deer and wolf over the ages.

Under climax conditions, the wolf population would also be low and characterized by an old-age structure and low recruitment. It would exist at the edge of subsistence, with the population level set by the availability of vulnerable prey. Adults would maintain themselves even though resources for reproduction would be rendered unavailable (low P/B ratio), as reported for one wolf pack in Minnesota by Mech (1975a, 1977b).

Minor shifts in equilibrium of this hypothetical pristine state occur because of periodic, normal variation in the environment. Suppose, for example, an abundant acorn crop occurred. Resources that normally were in the canopy of forest trees beyond the reach of deer would rain down like manna. The deer herd would shift from a diet of low quality to one of exceptional quality. Subsequently, the conception rate would increase. In the following winter, many older deer that would ordinarily have been vulnerable to predation would manage to escape, due to their improved condition. The wolf population, already close to subsistence, would suffer hard times with the reduced number of substandard deer, and the more vulnerable individuals would fail to survive.

In the following season, vegetation growth (which is independent of the acorn crop) would occur as usual, but the increased size of the deer herd would put more pressure on the preferred vegetation. To get sufficient quantity, less palatable parts of favored species and less palatable species would have to be consumed in greater quantity. The average quality of deer diets would decline, as would the condition of the deer. The higher than normal young-of-the-year crop would suffer first and most drastically, followed by the decline in condition of older animals. Conditions for wolves would be improved with the availability of a plentiful, weakened young-of-the-year class and the year's carry-over of vulnerable old animals. Predation would begin to reduce the level of the deer herd, thereby easing the pressure on the vegetation. Given no further periodic events, the system would return to the equilibrium state. However, time lags often result in oscillations in these responses. The sequence illustrates how a factor that influences directly one point in the system (food for deer) has ramifications throughout the system.

Now, consider a second example—the effects of an exceptionally hard winter. Deep snows would hamper the movements of deer and reduce food availability, which would increase maintenance requirements. Vulnerability to wolves would increase. The food base of the wolf population would increase and the deer herd would be reduced. The following spring, deer could be more selective of food quality because of lowered intraspecific competition. The birth rate might be reduced, but the survivorship of young produced would be high, and the average health of adults would increase (although there may be a year time lag if effects of the hard winter carry over into summer). Under reduced browsing pressure, plant biomass would increase. Because of the improved condition of deer, vulnerability to wolves would decline and recovery would begin. In this case, the deer compartment would decrease in size during disturbance, then increase during recovery.

Numerous other possible periodic events will occur to the reader, including simultaneous events. However, the reasoning of the conceptual model should be made clear by the examples. In fact, in the real world such periodic events occur with such frequency as to obscure the dynamic equilibrium (except over broad areas or over long time periods).

Turning from the pristine to the more typical, man-influenced present, we can consider the implications of the model in cases of rather severe and continuous disturbance by man. Consider, first, the case of artificial removal of large predators from most of the range of deer (fig. 1.1*B*). The deer population would grow and become older in age structure. Recruitment would decline to a very low level, and the P/B ratio would be extremely low. Essentially, the deer would be at the subsistence level. Vegetation would receive extreme browsing pressure. Standing crop biomass of favored plant species would be reduced and P/B ratios would be the maximum that the vegetation could sustain.

Achieving an equilibrium under these conditions would be unlikely because of periodic events, such as hard winters. Severe damage of the vegetation and die-off of deer would be expected. In environments with pronounced seasonality, extreme oscillation would be likely. In more stable, year-long climates, a subsistence equilibrium might be achieved. Man could (and frequently has) stabilize such situations by the substitution of human hunting for natural predation.

Consider a second case in which fire is used to create successional stages of vegetation favorable to deer (fig. 1.1*C*). This approach was commonly used by American Indians to encourage game but was not accompanied by predator control. Burning would reduce much of the vegetation biomass that was previously out of the reach of deer. Thus, the total standing crop would be reduced, but the proportion that was of good quality and within reach of deer would increase. Under such food conditions, the deer population would increase. Because of good health, deer would be less vulnerable to wolf predation. Also, harvest levels by Indians appeared to have been quite light (Elder, 1965; B. D. Smith, 1975). As subsistence was approached, P/B ratios of deer would decline and recruitment fall. The impact of heavy browsing would be to slow the successional rate of vegetation by keeping it in suitable deer range for a longer time, thereby maintaining relatively high P/B ratios.

Predators, though they would have a hard time during the deer population buildup because of low vulnerability, would achieve good populations based upon an abundant deer herd with an old-age structure. A dynamic equilibrium would be possible but unlikely because of periodic events. Each disturbance resulting in a lowered deer-browsing pressure would allow an irreversible, successional trend, with more of the woody vegetation escaping from deer reach. Therefore, as some plants escape suppression, the standing crop biomass would increase and the average P/B ratio of the vegetation would decline. The competitive advantage and shading effects of larger trees would continue to reduce the quality of the shrub and herbaceous layer for deer, even if further plants did not escape. Thus, an equilibrium would be unlikely to persist because periodic events would favor maturation of the vegetation and return to pristine condition. The trend would have to be countered by periodic setting back of succession by man.

What would happen if both disturbance of climax vegetation and predator control occurred simultaneously? The removal of predators from favorable deer habitat would lead to an even greater instability than in pristine environments. Greater extremes in oscillation would be expected since the capacity of growth of the deer population would be greatly increased. This describes the "irruption" phenomenon so common several decades ago (Leopold, 1943; Leopold et al., 1947; Taylor, 1956; Caughley, 1970).

A final example is the typical case in the northern Great Lakes region in recent years. The habitat has been gradually reverting to mature forest. Natural predators have been effectively eliminated. Hunter kills of deer have been moderate to light because of resistance to antlerless deer hunting. Deer populations usually are near subsistence level, have old-age structures and poor recruitment, and are subject to periodic winter losses. What would happen if maximum sustained hunter kill could be obtained?

According to the model, the increased kill would lower the size of the deer herd but increase recruitment. The P/B ratio would increase as the standing crop biomass was reduced. Winter die-off would be less probable as better health of deer would allow them to go into winter with greater fat reserves, and winter density would be lower. This, in turn, would lessen the browsing pressure on the favored food. Plant biomass would increase and the P/B ratio of the vegetation would decline. The rate of succession to mature forest would increase.

The George Reserve Experiment

The George Reserve deer population has been experimentally manipulated to test this ecosystem hypothesis. Because the reserve is a natural area, it was not possible to manipulate the vegetation directly. Therefore, the vegetation was manipulated indirectly by varying the size of harvest of the deer population. If the hypothesis was correct, the vegetation should show responses to the variation in the size of the deer population. In essence, man played the role of the large predator, albeit without the selectivity shown by natural predators. Once the responses of the system were known, the effect of predator selectivity could be simulated.

The strategy was to alter the level of the deer population in discrete steps. Once a change had been made, the deer population responses and the vegetation responses were measured. When I assumed responsibility for managing the population in 1966, it was being held to a post-hunt population of eighty as determined by the drive count (see chap. 2). Therefore, the population was maintained at a drive count population of eighty for three years, then reduced to sixty for four years and forty for two years.

Modeling the deer population on the basis of previous data suggested that time-lag effects for adjustments of deer population and vegetation responses would be minimal at these low densities. That is, the manipulation of the harvest in one fall and winter would show responses in the following summer.

It was apparent from the population model that the effect of further reductions below forty could be modeled without performing the actual reductions, because extrapolation was minor in comparison to the range of values already obtained and such extrapolations were constrained by having to pass through the zero intercept at total elimination of deer.

The actual data obtained from this series of step reductions based on the drive count did not show corresponding steps in the actual population; rather, a fairly consistent decline in the population was observed (see fig. 3.1). This result occurred because of error in the drive counts (see chap. 2). However, because time-lag problems were not involved, the desired objective of determining the deer population and vegetation responses in relation to declining deer density was achieved.

In 1973–74 an attempt was made to complete the reconstructed population and end the reduction phase of the experiments by shooting all of the deer that were older than fawns. In 1974–75 this attempt was continued, but was restricted to yearling and older animals. If all of the older animals could have been removed, then only the fawn crop would have had to be predicted from the deer population model. Subsequent analysis of the kill data and reconstructed herd showed that the attempt to remove all adult animals had failed. From field observations it was known that some adults had been missed, and we hoped that this number would be trivial. Unfortunately, the number appeared to be substantial. The effort substantially reduced the population to below forty.

The data on the final reduction step of the experiment (1971–72 to 1974–75) will not be available until all of these animals are dead—which may be as long as twelve years. Therefore, it seemed undesirable to further delay the presentation of the results obtained thus far. Data included in this paper cover nineteen years (1952–53 through 1970–71). The remaining data on the reduction phase will be published when they are available.

Completion of the entire set of experiments requires allowing the deer herd to build up to high densities to test the validity of extrapolations of present data in that direction. This phase of the experiment, which was begun in 1975, will require a much longer time because of time-lag effects.

CHAPTER 2

Methods of Study and Management

Drive Counts

Considerable variation occurred in the conduct of the drive census in the early years. Jenkins (1964) has reviewed this topic, and it will not be repeated here. Since 1946 the same basic format has been used. Drivers line up along the east fence and move across the reserve, east to west, until they arrive at the west fence. Deer forced to pass back through the line are tallied by the driver if they pass through on his right. Deer occasionally pass through the line from behind and the driver on whose right they passed reports a negative number. At the termination of the drive, numbered cards distributed at the beginning are collected and the total tally is determined. This number is termed the *drive census*. The number of deer found dead, poached, and harvested prior to the drive are added to the drive census. This total constitutes the *drive count*.

There are two basic sources of errors in the drive census: those due to the behavior of deer, and those due to the people in the drive line. Neither is easily overcome. The behavior of deer varies substantially from day to day. If the deer move out ahead of the drive line, coalesce into large groups, and rush back through the line, a good count will be obtained. If the deer do not move out ahead of the line, remain as singles, pairs, or small family groups, hide by lying down, or sneak back past the drivers, a poor count will be obtained. Similarly, in a drive census in Minnesota, Tester and Heezen (1965) reported that one telemetered deer moved a mile outside of its normal range while another remained within its range during the drive.

I do not know the cause of the difference in deer behavior. From experience with driving crews trying to push deer past shooters on stands, it was apparent that weather conditions profoundly influenced deer behavior. On some days the drive crew would begin to move and shortly thereafter the shots would begin a half mile to a mile ahead of the line. On other days no deer would be seen by either drive line or shooters. On still other occasions, the deer would be seen but would refuse to be moved in the desired direction. They would charge back through the line despite our best attempts to cut them off and turn them back. The relationship between deer behavior and weather is not a simple one. The study of Newhouse (1973) identified some factors involved, but the interaction effects were extremely complex and predictability of behavior was low.

There is extensive brushy cover on the reserve, particularly in the large tamarack swamps where thick tangles grow (see fig. I.2). If deer lie down in heavy brush, the probability of seeing them is extremely low. Many a driver has been badly shaken by a deer bolting from cover a few meters away. On one occasion a group of about thirty-five observers was waiting along the west fence for the drive line when a large buck came running along the fence. Less than fifty meters away, in clear view of the group, the buck moved into a small tangle of brush and bedded down with its head on the ground in the fawn hiding posture. After a few minutes it got up and ran away.

Two changes were made in drive procedures in an attempt to reduce the problem of deer hiding and sneaking. The first was to increase the number of people on the drive line and to concentrate them in the areas of heavy cover. (The relationship of number of drivers to the resultant count is discussed later in this chapter.) The second change, initiated in 1967–68, involved shifting from the traditional, early December drive date to early January. Visibility was increased due to the loss of persistent and/or dried leaves and the greater probability of a snow cover to press down herbaceous vegetation and form a contrasting background.

Many of the changes in the drive were made to solve organizational problems of coordinating the movement of a large number of volunteer people who are inexperienced and untrained. The largest single problem is keeping the line straight with the people correctly spaced. In the rough terrain and heavy cover, visibility is poor; if any two drivers lose contact with each other, a gap can develop. It is in heavy cover that deer are most likely to be missed, as they attempt to sneak through such gaps in the drive line.

In 1966 eight, east-west guide lines were marked by applying paint spots to trees on the reserve. In addition, two north-south lines were painted—one near the beginning of the drive and one near the end. Captains were placed on each of the eight east-west lines to coordinate movement in that direction. The entire drive line was stopped at the two north-south lines and reorganized before proceeding. The switch from December to January resulted in a major reduction in the problem of gaps forming in the line. Prior to that time, the drive line had to split to avoid numerous ponds, marshes, and wet places in swamps and bogs. Reestablishing connections was difficult and uncertain. By January virtually all water on the reserve is covered by ice, and the line can pass straight across.

Other human errors come from duplicate recording of the same group of deer, failure to record, etc. Because the people on the drive line are volunteers, most are there to have a good time and some are not very concerned about the outcome. Motivating and organizing this many people is possible under favorable conditions. However, southern Michigan winters are not famous for being favorable. For example, in 1974–75 it rained throughout the drive. Deep snows occurred other years. The worst conditions were in 1971–72, when the temperature was −13°F and there was a strong wind from the west, the direction of the drive. Under unfavorable conditions many drivers become miserable and exhausted. They simply want to finish the drive and are not concerned about the count of deer. In recent years we have checked each driver through a gate at the end of the drive and tried to detect errors in double recording or failure to tally animals.

Reconstructed Population

Jenkins (1964) was the first to use the reconstructed method of determining the size of the population in deer. Hesselton, et al. (1965) used the same approach with deer in New York, and more recently Lowe (1969) attempted the same method with Scottish red deer (*Cervus elaphus*). This general approach to population estimation has been used by fisheries workers for many years and is known as the *virtual population estimate* or *biostatistical method* of population analysis (see Ricker, 1975, for a review). This approach is possible on the reserve because virtually all of the deer are eventually accounted for in the harvest or are found dead. Lower jaws were collected and saved since 1952. Based on age determination by tooth replacement and wear, Jenkins was able to reconstruct the composition of the population at some earlier time. Thus, if a 6.5-year-old deer was shot in 1960, it had to have been born in 1954 and was in the population as a year older animal in each subsequent year until its death. After all of the deer alive in 1954 have died and have been assigned ages, it is possible to add the total to get a minimum number present in 1954. When Jenkins did this for the collection of jaws, he found that the reconstructed population exceeded the drive count for the years covered. Since the reconstructed population was a minimum, he concluded that this estimate was more relieable than the drive count.

The reconstructed population is the basic data set used in this study. It is far more reliable than the drive count. The problem with it, however, is that the reconstruction of the population cannot be completed for a given year until every deer alive in that year has died. Since the oldest deer age recorded at the reserve was 12.5 years, the reconstructed population cannot be completed until twelve years after the recording date. Relatively few deer live that long, and one can usually live with errors introduced by extreme cases of longevity. Nevertheless, there is a long lag between the present and the most recent year in which the reconstructed population is close to completion. It is some comfort to know that eventually the reconstruction will be completed and the population known. Still, for current management decisions about the harvest, it is necessary to use the drive count, even though that count is subject to considerable error.

The reliability of the reconstructed population is dependent upon two factors: (1) that the ages assigned to the jaws are accurate, and (2) that all of the deer, whatever the time and cause of death, be accounted for.

Determining the age of deer by tooth replacement for fawns (young of the year; 0.5 years; 0 age-class) and yearlings (1.5 years; 1 age-class) presents no problem, and the results are completely reliable. Even very large yearlings shot late in the spring showed obvious signs of recent replacement of milk premolars, but these were double-checked by cementum layer counts. For 2.5-year-old and older deer, Jenkins (1964) relied upon amount and pattern of tooth wear according to the method of Severinghaus (1949), as modified by the extensive study of Ryel et al. (1961) in which specific age criteria were developed for Michigan deer. These criteria were used by Jenkins and the state of Michigan Department of Natural Resources deer specialists to arrive at estimations of age for the deer from the reserve.

Subsequently, Bromley (1968) determined the age of these deer using the more reliable cementum layer technique on the first permanent molar—a modification of the general methods of Ransom (1966) and Mitchell (1967). A considerable number of errors in age assignment by wear were found, and these increased with the age of the deer. Similar results were reported for white-tailed deer by Gilbert and Stolt (1970) and Lockard (1972), for mule deer (*Odocoileus heminous hemionus*) by Erickson et al. (1970), and black-tailed deer (*O. h. columbianus*) by Thomas and Bandy (1975).

The cementum layer technique, modified to shorten the handling process, has been used to determine the age of all deer taken since Bromley's work. The technique involves sectioning the first permanent molar, grinding and polishing the thickened cementum pad that lies on the bottom of the tooth between the two roots, and examining it under a disecting microscope with reflected light. The growth annuli can be counted as darker lines or layers in the cementum accumulation. By alternately grinding and examining, usually some section can be found that shows the layers clearly. If questions of age determination arise (not all specimens show the layers equally clearly), the other first molar or additional molars are examined.

Thus, the results reported by Jenkins and those for the same period given here will not be in complete agreement. Furthermore, a complete recheck of the original records presented by Jenkins showed discrepancies. These discrepancies have been corrected or resolved where possible, but where they could not be resolved, the figure reported by Jenkins was used.

The second factor—accounting for all of the deer—involves both the problem of animals being missed and the problem of what to do about animals accounted for, but for which age and/or sex are unknown. The latter include escapees during deer drives held prior to the raising of the fence and animals that were poached.

The problem of animals dying in the reserve and not being found is particularly intractable since it is difficult to cross-check. The works of Verme (1962) and Langenau and Lerg (1976) in Michigan, Teer et al. (1965) in Texas, and O'Pezio (1978) in New York have emphasized the postparturition period as a time of much of the mortality of young. Clearly, small dead fawns would be easily missed. Furthermore, foxes, raccoons, opposums, etc., could readily devour completely such small carcasses with low mineralization of bones. Few early mortalities of fawns are recorded for the George Reserve.

Unfortunately, it has never been possible to determine directly the number of fawns born on the reserve. Crawford (1957) made considerable effort to capture fawns and managed to capture only four, while Queal (1962) captured six. My assistants and I have made attempts and captured a maximum of five in a given year.

The problem is circumvented in this study by using the number of young recruited rather than the number of young born. Recruitment is defined as the number of young of the year alive when harvesting is carried out. Since most fawns are born in June, and most animals are harvested in December and January, six months would be the average age of recruitment. Over the period of this study, virtually no natural mortalities occurred after recruitment age. Every case of poor condition, either in animals harvested or found dead, has been due to injuries (nearly all gunshot wounds). Most deer on the reserve die violent deaths.

When one compares the recruitment rate with the conception rate (see chap. 4), it is obvious that at higher population densities some fawns must be dying before recruitment age. If one compares the number recruited to the number expected from embryo counts at the lower population densities, the numbers are virtually identical. Thus, early fawn mortality is pronounced at higher densities, but negligible at lower densities.

By recruitement age, the carcasses of fawns are as probable of being located as those of older animals. This is particularly true when one takes into account the clues to location which are independent of carcass size: presence of avian scavengers, density of terrestrial scavenger scats, smell of rotten carcass, and scattered bones. Hence, finding carcasses of fawns of recruitment age can be treated as part of the larger problem—are carcasses of older animals being missed and decomposing without being recorded?

It is my opinion that nearly all carcasses of older deer are found. The George Reserve is only about two square miles (4.64 km^2) in extent and is a heavily used field research area. There are seven living accommodations on the reserve for research workers that are occupied much of the time. Recreational walking by resident researchers and spouses is common. There is a resident caretaker and his three-man crew who spend part of their time working on the area. Personnel with the deer research project spend considerable time looking for dead deer. People on the drive census are instructed to report dead deer. Deer bones persist for a minimum of two and up to four or more years depending upon the site. They would most likely be missed in the tamarack swamps, where relatively fewer people go; however, this is where the greatest concentration of people is placed during the drive censuses. In short, the probability of carcasses being missed is relatively small.

A second, rough cross-check of the accounting for animals can be made by using the animals marked by Hirth (1977) from 1968 to 1970. A total of forty-five deer were marked with color-coded ear tags. In addition to ear tags, yearling and older animals were fitted with collars constructed from heavy, rubber-impregnated canvas drive belting, strongly bolted together at the botton of the loop. A number of ear tags were lost, particularly from animals tagged as fawns. Many recovered tags (both found in the field and on animals shot) were heavily dented by deer teeth marks. The tags must have been chewed on by other deer, since the tag carrier could not possibly have reached them. This factor probably accounts for many of the lost tags. No case of a collar breaking was known, but at least several smaller deer managed to slip the collar over their heads. Thus, loss of markers confuses the data, particularly in the case of fawns carrying only ear tags.

Of the forty-five marked deer, twelve were unaccounted for, three were found dead, and thirty were killed in the harvest. Of the twelve unaccounted for, four were marked as newborn or very young fawns and it is doubtful that they lived to recruitment age. Of the remaining eight, it is almost certain that some are still living. Four were fawns of recruitment age which were marked (ear tags only) in 1970, as were a yearling and a two-year old. The remaining deer were tagged in 1968 and 1969, mostly as young adults. There is a good probability that some of the unaccounted marked animals were poached. Two of the three marked animals found dead died of poacher gunshot wounds. Cause of death in the third deer could not be determined, but it was in excellent condition and shooting is highly likely.

Also, six out of seven marked animals from the study of Queal (1962) were killed or found dead. Loss of ear tags on the other individual is a distinct possibility, since many of the original ear tags and all of the plastic collars were eventually lost from Queal's animals. Eventually, marked animals still living will be accounted for and a more complete assessment made, but from the evidence available at present, it appears that most animals of recruitment age or older that die are being found.

Deer are poached at night with a spotlight from vehicles on roads that run along parts of the two sides of the reserve. The animal is shot, a hole is cut in the fence, and the deer dragged out.

A fire lane, approximately eight-feet wide, is plowed along the inside of the fence. By patrolling the fence regularly, holes can be discovered and drag marks and other signs of removal of deer examined. Thus, the record of deer poached is quite good, but sex and age are almost never known. Similar problems pertain to early drive census escapees, whose sex was often determined, but not age. There are also a few gaps in the early record where a deer was recorded, but sex and age information was lost. The total number of such cases is 55 for the nineteen years of reconstructed population data, or 6.6 percent of the total removal of 826 animals.

Omitting these animals from the reconstructed population would result in a large underestimation since it is sure that they were a minimum of 0.5 years old. However, entering them as 0.5-year-olds will also be an underestimation since it is also sure that at least some of them were older animals. An underestimate of age would lead to animals being entered into the totals of those of fewer years, with the consequence of lowering (by omission) the total for years when the animal was alive but not included. On the other hand, if they are entered at ages greater than their actual age, the results will be biased toward the high side.

The approach taken was to examine the sex and age composition of the combined total of deer found dead from gunshot wounds, including harvested animals that carried previous gunshot wounds. It was assumed that the sex and age composition of deer wounded by poachers but escaping to die later or be shot by us was similar to the composition of deer poached and removed from the reserve. A total of thirty-one deer that had been wounded by poachers included twenty males and eleven females. It was concluded that, whether by deer vulnerability or poacher selectivity, animals shot by poachers were predominately males. Hence, animals of unknown sex removed from the reserve by poachers were allocated to sex by the ratio two males to each female.

The age composition of male deer wounded by poachers was comprised of two fawns, fifteen yearlings, and one each of two-, three-, and four-year-olds. Since fawns are far more prevalent in the population than yearlings, they must be either less vulnerable or are not selected by poachers, as is true of the harvest (see fig. 5.2). That fawns may not be selected was suggested by their total absence in the female kill. Because of the overwhelming preponderance of yearlings in males wounded by poachers, this was the age class into which males were entered for unknown sex and age animals poached and removed. The two fawns and three older males nearly balance the age bias, but entering males at high ages would bias the population data more, because the entry would be made in more years. Therefore, the yearling age was used so the bias in the reconstructed population would be on the conservative side.

The age composition of the eleven females wounded by poachers was six yearlings, two 2.5-year-olds, and three 3.5-year-olds. Females were equally allocated to the unknown poacher removals as yearlings and 2.5-year-olds. The procedure still contains a bias toward the low side, but not as low as entering all of the unknowns as fawns. It seems like the best solution under the circumstances.

Harvest Methods

Prior to 1953–54 the harvesting of deer was done by the resident caretaker and a few, select university personnel. From 1954–55 until 1968–69, harvesting was done by a drive method with a crew of beaters pushing deer past shooters on stands. High-powered rifles were used to do the killing (shotguns with slugs or muzzle-loaders are required in the general deer season in southern Michigan) and shooters were placed so as to shoot against the hillside as a backstop. Deer were to be shot either in the neck or the heart. No shot was to be taken at distances greater that 75 yards (69 m), and only standing or slowly moving deer were to be shot.

A volunteer crew of approximately eight to twenty people was used to move deer past the shooters. Usually, two or three drives were made in a day. After each drive, a pickup truck went around to the shooters and transported the whole, freshly killed deer to the dressing station for weighing, processing, and collection of biological samples. This method required a tremendous amount of manpower and coordination for the number of deer harvested. Some of the shooters were competitive, particularly regarding the taking of bucks. Selection of bucks during this period was apparent (see chap. 5). Either regulations governing shooting were not adhered to, or the competence of some shooters was low, since many poorly shot animals arrived at the dressing station and a considerable amount of time was spent tracking wounded deer.

Toward the end of the 1968–69 harvest, the use of a spotlight and shooting at night from a pickup truck was tried. This method proved so successful that it has been used since that time, although the period of shooting was expanded to include dawn and dusk as well. The low manpower requirement in relation to the number of deer killed and clean kills characterized the method. Anticipating a public reaction to the use of a method favored by poachers, we issued news releases to explain the reasons for this course of action. Not a single response, pro or con, was received. Our public relations appeared to be much improved with the passing of the image of a private hunting club.

CHAPTER 3

Population Size

The reconstructed population is the most reliable estimate of the true population of deer on the George Reserve. Because any error in accounting for all of the dying deer lowers the total, this estimate is on the conservative side but approaches closely a total count.

Data on prehunt numbers, size of kill, and posthunt numbers are given for males and females in tables 3.1 and 3.2, respectively. Kill figures represent the actual data, while prehunt numbers represent the reconstructed herd. Posthunt numbers were derived by subtracting the kill from the prehunt numbers. Therefore, numbers in the categories are entirely consistent with each other. It should be noted that the data are not derived from a sample, but rather are a close approximation of a complete count. Thus, the concern in analysis is not "how good is the estimate?" but "what can be concluded from what was observed to have happened?"

The terms posthunt and prehunt are based upon the conventions of big game hunting seasons in North America. The conventional terms are used since they reflect the sequence used on the George Reserve deer herd. One might more precisely state that posthunt population is that population engaging in reproductive effort, and prehunt population is the total population size at the time surviving young are designated as recruits.

It is convenient in this treatment to consider the total population of the reserve. It is common for studies of deer to express values as densities, most commonly as numbers per square mile. The reader may derive quick approximations of the numbers per square mile for these results by halving the total for the reserve, since the area is approximately two square miles. More precise conversions can be obtained by multiplying the total for the reserve by 0.559 to get deer per square mile, or by 0.216 to get deer per square kilometer.

During the nineteen-year period for which the reconstructed

TABLE 3.1. Number of Males by Age Class in the Prehunt and Posthunt Population and the Kill for Nineteen Years

Age Class	*Year* 52-53	53-54	54-55	55-56	56-57	57-58	58-59	59-60	60-61	61-62	62-63	63-64	64-65	65-66	66-67	67-68	68-69	69-70	70-71	Total	Percent
Prehunt																					
0	13	19	34	24	40	22	25	23	23	20	12	22	26	21	22	25	26	24	18	439	44.0
1	16	12	17	28	23	22	17	22	19	15	9	11	13	19	13	10	15	15	11	307	30.8
2	11	7	5	7	6	6	4	7	9	3	6	5	4	5	8	4	5	4	7	113	11.3
3	4	11	4	4	4	4	3	3	3	8	3	6	4	2	3	5	3	1	2	77	7.7
4	3	4	4	2	1		2	2	1	2		3	1	2	1	2	3			33	3.3
5		3	1	1										1	1			2		9	0.9
6			2	1											1	1			1	6	0.6
7				2												1	1			4	0.4
8																	1	1		2	0.2
9	1																	1	1	3	0.3
10		1																		1	0.1
11	1		1																	2	0.2
12				1																1	0.1
Total	49	57	68	70	74	54	51	57	55	48	30	47	48	50	49	48	54	48	40	997	99.9
Kill																					
0	1	2	6	1	18	5	3	4	8	11	1	9	7	8	12	10	11	13	10	140	30.8
1	9	7	10	22	17	18	10	13	16	9	4	7	8	11	9	5	11	8	6	200	44.1
2		3	1	3	2	3	1	4	1			1	2	2	3	1	4	2	1	34	7.5
3		7	2	3	4	2	1	2	1	8		5	2	1	1	2	3	1	1	46	10.1
4		3	3	2	1		2	2	1	2		3		1	1	2	1			24	5.3
5		1		1														1		3	0.7
6				1															1	2	0.4

Age Class	Year 52-53	53-54	54-55	55-56	56-57	57-58	58-59	59-60	60-61	61-62	62-63	63-64	64-65	65-66	66-67	67-68	68-69	69-70	70-71	Total	Percent
7				2																2	0.4
8																				0	0.0
9																		1		1	0.2
10	1																			1	0.2
11																				0	0.0
12				1																1	0.2
Total	11	23	22	36	42	28	17	25	27	30	5	25	19	23	26	20	30	26	19	454	99.9
Posthunt																					
0	12	17	28	23	22	17	22	19	15	9	11	13	19	13	10	15	15	11	8	299	55.1
1	7	5	7	6	6	4	7	9	3	6	5	4	5	8	4	5	4	7	5	107	19.7
2	11	4	4	4	4	3	3	3	8	3	6	4	2	3	5	3	1	2	6	79	14.5
3	4	4	2	1		2	2	1	2		3	1	2	1	2	3			1	31	5.7
4	3	1	1										1	1			2			9	1.7
5		2	1											1	1			1		6	1.1
6			2												1	1				4	0.7
7																1	1			2	0.4
8																	1	1		2	0.4
9	1																		1	2	0.4
10		1																		1	0.2
11			1																	1	0.2
12																				0	0.0
Total	38	34	46	34	32	26	34	32	28	18	25	22	29	27	23	28	24	22	21	543	100.1

TABLE 3.2. Number of Females by Age Class in the Prehunt and Posthunt Population and the Kill for Nineteen Years

Age	*Year*																				
Class	*52-53*	*53-54*	*54-55*	*55-56*	*56-57*	*57-58*	*58-59*	*59-60*	*60-61*	*61-62*	*62-63*	*63-64*	*64-65*	*65-66*	*66-67*	*67-68*	*68-69*	*69-70*	*70-71*	*Total*	*Percent*
Prehunt																					
0	14	29	15	23	28	16	12	15	18	20	19	26	23	19	23	18	25	17	23	383	34.4
1	6	13	24	11	23	18	10	12	8	14	14	17	14	21	14	20	10	13	11	273	24.6
2	7	4	7	15	11	9	10	9	4	5	6	12	9	13	15	9	16	8	7	176	15.8
3	7	7	3	5	12	3	5	9	5	3	3	5	7	5	9	10	7	5	4	114	10.3
4		7	5	1	5	6		3	3	5	1	1	2	6	4	6	4	2	3	64	5.8
5	2		4	2	1	3	1		3	3	1	1	1	1	5	3	2	3		36	3.2
6	2	2		1	2	1	1			3	2	1	1	1		5	3	2	1	28	2.5
7		2			1	2	1	1			1	1	1	1	1		4	3	2	21	1.9
8			1					1				1	1		1			2	2	9	0.8
9				1									1			1			1	4	0.4
10					1									1						2	0.2
11															1					1	0.1
12																1				1	0.1
Total	38	64	59	59	84	58	40	50	41	53	47	65	60	68	73	73	71	55	54	1112	100.1
Kill																					
0	1	5	4	1	10	6		7	4	6	2	12	2	5	3	8	12	6	12	106	28.0
1	2	6	9		14	8	1	8	3	8	2	8	1	6	5	4	2	6	5	98	23.9
2		1	2	3	8	4	1	4	1	2	1	5	4	4	5	2	11	4	2	64	16.9
3		2	2		6	3	2	6		2	2	3	1	1	3	6	5	2	2	48	12.7
4		3	3		2	5				4			1	1	1	4	1	2		27	7.1
5			3			2	1			1				1				2		10	2.6
6		2								2	1					1				6	1.6

Age Class	Year 52-53	53-54	54-55	55-56	56-57	57-58	58-59	59-60	60-61	61-62	62-63	63-64	64-65	65-66	66-67	67-68	68-69	69-70	70-71	Total	Percent
7		1			1	2		1					1		1		2	1	1	11	2.9
8								1					1					1	2	5	1.3
9																1				1	0.3
10					1															1	0.3
11																				0	0.0
12																1				1	0.3
Total	3	20	23	4	42	30	5	27	8	25	8	28	11	18	18	27	33	24	24	378	99.9
Posthunt																					
0	13	24	11	22	18	10	12	8	14	14	17	14	21	14	20	10	13	11	11	277	37.7
1	4	7	15	11	9	10	9	4	5	6	12	9	13	15	9	16	8	7	6	175	23.8
2	7	3	5	12	3	5	9	5	3	3	5	7	5	9	10	7	5	4	5	112	15.3
3	7	5	1	5	6		3	3	5	1	1	2	6	4	6	4	2	3	2	66	9.0
4		4	2	1	3	1		3	3	1	1	1	1	5	3	2	3		3	37	5.0
5	2		1	2	1	1			3	2	1	1	1		5	3	2	1		26	3.5
6	2			1	2	1	1			1	1	1	1	1		4	3	2	1	22	3.0
7		1					1				1	1		1			2	2	1	10	1.4
8			1									1			1			1		4	0.5
9				1									1						1	3	0.4
10														1						1	0.1
11															1					1	0.1
12																				0	0.0
Total	35	44	36	55	42	28	35	23	33	28	39	37	49	50	55	46	38	31	30	734	99.8

population is available, the prehunt (peak) population varied from a high of 158 in 1956–57 to a low of 77 in 1962–63 (fig. 3.1). Posthunt (subsequent breeding population) numbers varied from 89 in 1955–56 to 46 in 1961–62. Highest variation occurred in the earlier years. Attempts to stabilize the population from the middle 1960s and on were reasonably successful. However, the attempt since 1966 to reduce the herd in discrete steps was not accomplished. Rather, a smooth decrease of the posthunt population occurred.

Comparison of Drive Counts with Reconstructed Population Size

The reasons for doing a detailed comparison of the drive count and the reconstructed population are threefold. First, census methods for wild animals are notoriously poor, particularly those involving estimation of populations in heavy cover such as a hardwood forest. Any information on the variables which affect the results is valuable; indeed, improvements in census methods are dependent upon such information.

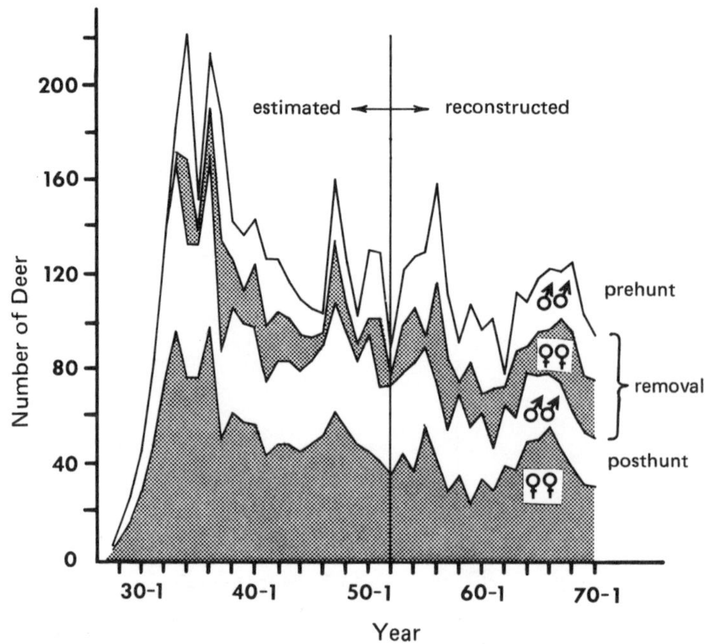

FIG. 3.1. Reconstructed and estimated population changes of white-tailed deer on the George Reserve

The drive count was the only estimate available for the nineteen years (1933–34 to 1951–52) prior to the collection of jaws—the basis of the reconstructed populations. Therefore, if a corrective equation could be derived to apply to these earlier drive results, interpretation of the early population changes could be more precise. Also, the time lag in obtaining the reconstructed population is so long that the drive count must be used to make yearly management decisions. Again, if a corrective equation could be developed to apply to the drive count, the desired experimental plan could be followed with greater accuracy.

Results of the drive count and the reconstructed population for the years when both were available are shown in fig. 3.2. From 1952–53 to 1956–57, the reconstructed population substantially exceeded the drive count. From 1957–58 to 1970–71 the two methods approached each other with variation in both directions. Since the reconstructed population is far more reliable, the drive counts contain most of the error. Particularly, the drive counts in 1963–64 and 1967–68 greatly exceed the maximum error possible in the reconstructed herd and must have been caused by duplication of tallies or false reporting of deer in the drive census.

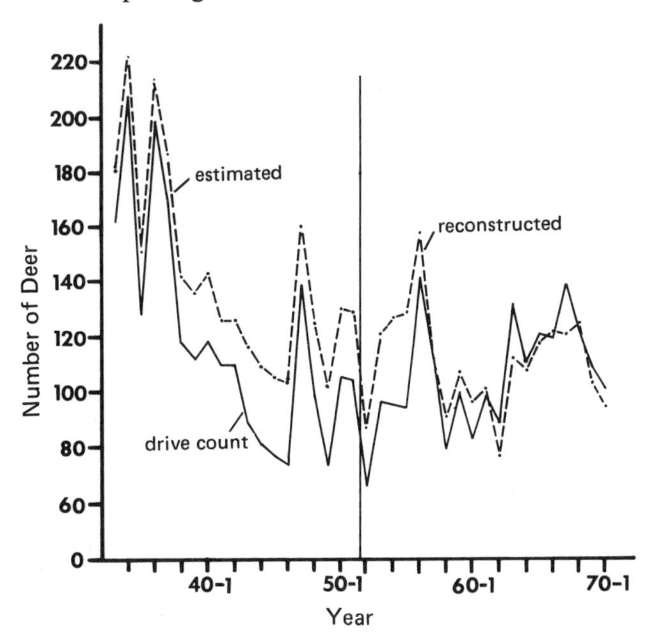

FIG. 3.2. Comparison of drive count and reconstructed deer population size

Correlations between the two methods are shown in fig. 3.3. Perfect correspondence is indicated by the 45° line. The two periods, one when the reconstructed herd greatly exceeded the drive count (1952–53 to 1956–57; $n = 5$) and the other when the methods approached agreement (1957–58 to 1970–71; $n = 14$), are treated separately.

The earlier period shows a good fit which approximately parallels the 45° perfect correspondence line. In view of the small sample size, it was considered essentially parallel (but not statistically significantly different from 1.0), and the drive count was almost consistently 21.5 percent lower than the reconstructed population.

For the latter period the points straddle the 45° correspondence line, but the slope is steeper. While the difference is not statistically significant at the 95 percent confidence level, it comes close (94 percent level). At high deer population levels the drive count overestimated the population, while at low populations it underestimated the numbers, thus explaining the difficulty we have had in obtaining reasonable drive counts in recent years when the population was very low. These results imply that at low populations the drive count is failing because of the stealthy behavior of the deer,

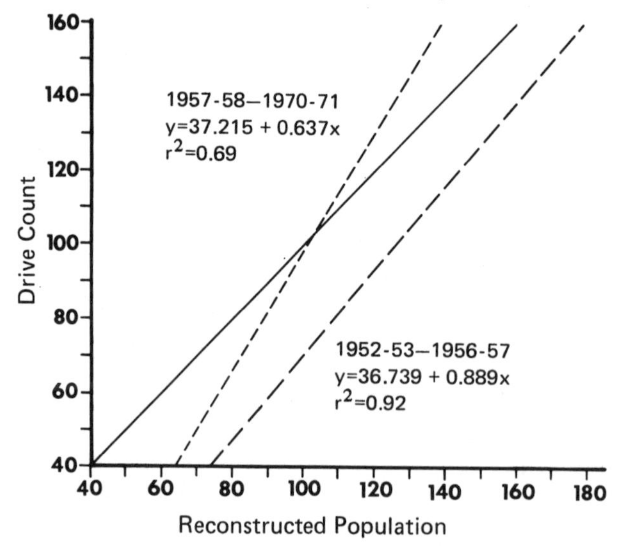

FIG. 3.3. Regressions of drive count on reconstructed population. Solid line is equality, broken line the five years of data, and dashed line the last fourteen years of data.

while at high populations the deer are being moved and observed but duplicate or erroneous reporting occurs.

This interpretation was supported by further analysis using multiple linear regression analysis on the combined data (n = 19). Simple linear regression of the drive counts on the reconstructed population size gave a coefficient of determination (r^2) of 0.46. Adding the number of people in the drive (range 50–144) to the multiple regression equation raised the R^2 to 0.58. Including date of the drive (which ranged from November 10 to January 11) had essentially no influence on the regression fit ($R^2 = 0.58$). Adding the year to the model, however, resulted in a substantial improvement of the fit ($R^2 = 0.81$). The year, of course, does not have biological reality, and its influence is attributable to improvements made in the organization of the drive line. Neither the number of drivers nor the drive date is independent of the year, and improvement in organization of the drive line was due to both.

Regressing the percent difference of the drive counts from the reconstructed population size on year gave an essentially linear fit with a positive slope ($b = 1.90$; $r^2 = 0.58$). The result highlights two difficulties of using the multiple regression equation to correct drive counts from the earlier years for which the reconstructed population sizes are not available. First, the assumption that the variance of the dependent variables is equal is probably not met. Variance seems to increase, being higher in the later years and lower in the earlier years of the nineteen-year sequence, as is also reflected in the relative fits of the two regressions in figure 3.3. Second, it is unlikely that the error was linear over the nineteen years prior to availability of reconstructed population sizes (1933 to 1951–52). In other words, the error could not be progressively worse (a 1.9 percent additional underestimation per year) as one moves back in time to 1933, or a negative error in 1933 of greater than 55 percent would result. Clearly linearity is unlikely, and the error must approach a maximum. Indeed, the regression for the first five years of reconstructed population data suggests that this maximum was approximately 21.5 percent.

In view of these considerations, it was concluded that the regression for the first five years of reconstructed population data (1952–53 to 1956–57) was the best equation for correcting the drive counts prior to the period when reconstructed population size was available (1933–34 to 1951–52). The population estimates shown in figures 3.1 and 3.2 were based on these regression corrections.

Infrared Scanning

In the winter of 1966–67 we attempted to count the population by the use of airborne infrared scanning. The infrared scanning determines the apparent temperature of the ground surface, and if the deer is warmer than the background, a "hot spot" is produced on the imagery. A count of such hot spots constitutes an estimate of the total population. Croon et al. (1968) described the technique in detail and presented the results of the full-scale trial on the George Reserve. McCullough et al. (1969) gave an assessment of the potential of the method for large mammal census.

Experienced interpreters reached a count of 93 positive and 5 probable deer from the infrared imagery for a total of 98. At the time, I estimated the actual population to be 101 (based on the drive count corrected by the regression of drive counts on reconstructed population sizes using data available up to that time). The reconstructed population size is now available for the year of the infrared count and a more complete assessment can be made. The reconstructed population was 122 and the drive count was 120. Since 40 animals were removed prior to the infrared count in January, 1967, the population at the time of the overflight was 82. Thus, the infrared count of 98 overestimated the number of deer. Apparently hot spots on the imagery were produced by objects other than deer.

Initial Population Growth

The increase of deer from 6 to 162 in six years intrigued early students of deer populations, since it exceeded the expectations based on what was known of the biology of deer at the time. O'Roke and Hammerstrom (1948) calculated a population projection assuming a population of 160 deer in 1933, and Jenkins (1964) projected 162, since 2 deer were found dead prior to the drive. Kelker (1947) calculated that this rate of increase slightly exceeded the expected rate based upon the breeding potential tables given by Leopold (1933).

Interesting questions are presented by this example of population growth. Did the population grow at an unimpeded rate over the six years, or were constraints to growth present before the end of the period? How does this initial population growth compare to population responses of recent years? Are they compatible or did the population behave differently with constraints arising at a later time? Derivation of the parameters necessary to answer these questions is explained in the following chapters. However, the end results will be used here to make the necessary comparisons.

In view of the fact that the drive count of 162 deer in 1933 was almost certainly low (the drive line consisted of about thirty people as compared with eighty to one hundred people in recent years), the corrected drive count estimate of 181 (fig. 3.2) is used.

Initial population can be derived from the number of animals introduced (2 adult males and 4 adult females that were presumed to be pregnant). Since the original stock was introduced on March 5 and March 13, 1928, the first reproductive period was the summer of 1928. Thus six reproductive seasons occurred between introduction and the first drive census on December 9, 1933 (Hickie, 1937). Negligible mortality of adults was assumed because of the low population densisty and the short time period involved; only the original adults might have died of old age. Sex ratio of offspring was assumed to be similar to that of recent years (see fig. 4.9). In calculations, all population numbers were rounded to the nearest whole animal.

The question of whether the population grew at an unimpeded rate can be tested by basing projections of population growth on the maximum rate of reproduction observed in this study and zero mortality. This projection given a population of 303 animals in six years and greatly exceeds the corrected drive count estimate of 181 deer. Therefore, the population reproduction statistics of recent years are more than sufficient to account for the observed growth, and some constraints must have been operating on the population during the six-year period. Changes in either reproduction rate or survival of fawns must have occurred to account for the lower total of 181.

If one assumes the same constraints on both reproductive rate and recruitment rate (i.e., survival of offspring) as have been operating in recent years, the projections give a population of 160 after six years. Obviously this is too great a constraint, and some time lag must have been present in the development of the constraint. It would appear that the constraint would be mainly in survivorship of offspring, since there is no reason to presume that reproductive rates would have been lower in the initial growth period than they have been in recent years. If rates were different, one would expect them to be higher rather than lower.

A decline in the reproductive rate with increasing population, equivalent to recent years, and no constraint on survival of offspring would result in a projection of 190 deer after six seasons. This number is quite close to the 181 estimated to be present. Thus, the reproductive rates observed in later years seem quite consistent with the initial growth of the population if one omits mortality of offspring or minor mortality (i.e., a nine-animal difference). Since both estimates contain some error factor, it is possible that there is no discrepancy.

It was concluded, therefore, that (1) this rate of growth is substantially below the maximum reproductive rates (at extremely low densities) of recent years; (2) the original growth of the herd was highly consistent with the realized reproductive rates in relation to density determined in recent years; and (3) constraints on survival of offspring did not occur until possibly the fifth or sixth year. Therefore, the projection plotted in figure 3.1 for the early population growth is based upon the realized reproductive rates of recent years for the first five seasons (up to 131 animals). The sixth year was constrained by 9 animals to correspond with the estimate of 181 animals derived from the corrected drive count.

Since the number of animals removed from the reserve by harvesting is known (O'Roke and Hammerstrom, 1948; Jenkins, 1964), and sex of most removals is also known, considerable information can be assembled on the early population history (i.e., before reconstructed population data are available) of the herd. Animals of unknown sex listed by these authors were assigned by sex as two males to one female on the same basis as the reconstructed population, since most of the unknowns were poached. Since sex ratio was not available for the posthunt population until the reconstructed population data began in 1952–53, the sex distribution of figure 3.1 for the early period is based on population projections until 1932–33 and the average posthunt sex ratio (57.5 percent female) of the reconstructed population for 1933–34 to 1951–52.

Trends in Population Size

The history of the George Reserve herd can be read from figure 3.1. Initially, following the introduction of six deer, the population grew at a rate consistent with recent reproductive performance. In the following several years the population remained high, but the rate of increase was offset by hunting removal. Expected increase and hunting removal, although showing discrepancies in some years, were reasonably consistent overall. Thus, population control during this period can be accounted for by hunting removal.

The period 1938–39 to 1946–47 was one of gradual reduction of the deer herd. Low recruitment and failure of the population to increase despite relatively low hunting removal correlated with reports of severe damage to the vegetation. Perhaps the vegetation had not yet recovered from the impact of earlier high populations. (This topic will discussed in greater detail in the section on time-lag effects in chapter 11.)

Considerable fluctuation in population size occurred between 1946–47 and 1966–67 when the present studies were begun. Peaks occurred (1947–48 and 1956–57), but the population was subsequently reduced by relatively heavy hunting. The overall trend in population, given the pronounced fluctuation, was downward, with a low reached in 1962–63. That was the year a threatened law suit over nonconformance of the reserve fence to Game Breeders License requirements halted the harvest. In the following few years, the population once again increased to moderate levels.

Beginning in 1966 an attempt was made to stabilize the fluctuations and institute a series of downward steps in the population level. As previously noted, stabilization was achieved, although the reduction took the form of a continuous cline rather than discrete steps.

CHAPTER 4

Reproduction

The rate at which female white-tailed deer conceive and bear young is a variable influenced primarily by quality of diet. This has been demonstrated repeatedly for *Odocoileus* (Morton and Cheatum, 1946; Cheatum and Severinghaus, 1950; Robinette et al., 1955; Taber and Dasmann, 1958; and many others). Under good conditions, the ratio of twins and triplets to singles increases, the number of females failing to conceive declines, and maturity is reached at an earlier age. Since Cheatum and Morton's (1942) pioneering study in New York, it has been known that white-tailed deer female fawns can reach sexual maturity in their first fall and bear young at slightly over one year of age. It should be noted that traditionally the term *fawn* means an animal in its first year, *yearling* the second year, and *adult* the third and later years of life. Adult does not refer to sexual maturity, since the females of all of these age categories are capable of reaching sexual maturity.

An assessment of the rate at which the George Reserve deer population produces new individuals is based upon data obtained from the examination of reproductive tracts of killed females. These examinations also provided information on the timing of the breeding season, the secondary sex ratio (at early embryonic state), the age-specific fecundity rate, and other related statistics.

The kinds and amount of reproductive data collected over the years has been variable, and direct comparisons among years are difficult. In the early days, caretaker Larry Camburn recorded the number of embryos found during the dressing of killed deer and recorded the sex of larger fetuses. These data, which extend from 1934–35 until 1954–55, allow a determination of the ratio of single to twin to triplet embryos. However, prior to 1951–52, ages of the females were not determined and the ratios according to age cannot be established. Reproductive information was not obtained in 1955–56 or between 1958–59 and 1962–63. Fortunately, reproductive tracts were examined in 1957–58 by Menzel (1958) and in 1963–64 by Dennis R. King. Tracts were collected in 1964–65 and 1965–66 by Drs. Chase and Cowan, and since 1966–67, I have collected them.

A major limitation of all the collections until recent years is that tracts were collected only from obviously gravid females. Females in early stages of pregnancy or those breeding late in the season were not included in the sample. Thus, while the embryos present represent a positive determination of pregnancy rate in the obviously pregnant females, it cannot be assumed that females not included were not pregnant. Indeed, it is certain that a sizable proportion were pregnant or destined to breed late in the year. Since 1968–69 all female tracts were collected, regardless of whether there were signs of pregnancy.

Ovaries of tracts collected since 1964–65 were examined for Graafian follicles and corpora lutea. Examinations were made macroscopically by the method of Cheatum (1949). Only structures of the current year were included because of the difficulty of correctly identifying older structures without histological preparation (Mansell, 1971; Hesselton and Jackson, 1974).

Time of Conception

Time of conception in females with detectable embryos was determined by using the age criteria of Armstrong (1950), principally the crown-rump and forehead-rump length curves. Armstrong gave data beginning with thirty-seven days of age, but younger embryos in this study were assigned ages based upon extrapolation of the obvious exponential curve back to the zero intercept. The youngest embryo that could be measured was estimated to be twenty-eight days of age.

The yearly time of conception of females since 1963–64 is presented in figure 4.1. Although the sample sizes were small, several patterns emerged. There was a tendency for yearling females to breed at a later date than the average for older females, and fawns were substantially later than older animals. The yearlings averaged only 1.4 days later than the mean for older females, while four fawns averaged 29 days later than older females, with a range of 14 to 37 days. These results agree with those of Haugen (1975) who found that in Iowa the average breeding date of yearlings was 2 days later than that of older females, and that fawn females peaked 11 to 26 days after older females.

A second point apparent from the figure is the shifting forward

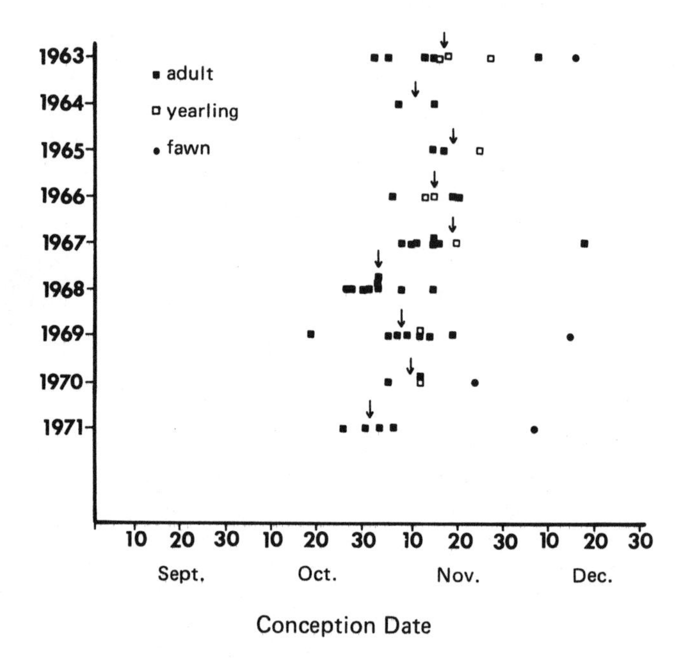

FIG. 4.1. Conception dates by year in George Reserve female deer as determined by backdating embryos. Arrows indicate mean conception date of yearling and older females.

of time of conception with the reduction of the population beginning in 1968–69. Between 1963–64 and 1967–68 the overwinter population was approximately seventy-five, while subsequently it was approximately fifty-five (see tables 3.1 and 3.2). The mean date of breeding for the five years with the higher population was November 16, with a range from November 11 to 19. After 1968–69 the mean conception date was November 6, with a range of November 1 to 10.

Verme (1963) found, from controlled feeding trials with white-tailed deer females, that diet influences the length and weight relationships of fetuses. However, these differences did not become apparent before 133 days of gestation. Nearly all the sample fetuses in this study were well below 133 days (only 3 out of 155 fetuses examined between 1963–64 and 1973–74 were older than 133 days), and the shift in density between the two periods could not have resulted in so great a difference in diet as between the excellent and poor diets of Verme's experiments. Therefore, while the variation in growth rate of fetuses could account for some of the difference, it is not sufficient to account for the magnitude of the shift. Thus, the time of breeding must be controlled to some extent by nutrition, as previously reported for white-tailed deer by Teer et al. (1965) and McGinnis and Downing (1977), and for tule elk by McCullough (1969).

Other Evidence of Pregnancy

On the George Reserve many females were killed before embryos were visible, and this was a particular problem with female fawns which bred about a month later than older females. However, changes in the female reproductive tract can aid in the determination of breeding activity of late breeding females. If the results from examination of reproductive tracts are plotted by date of kill (fig. 4.2) it is apparent that follicular activity predates conception by a considerable period of time. The earliest follicle activity was approximately sixty days before the mean breeding date of yearling and adult animals. These may have been animals that would have bred well ahead of the mean breeding date; however, the fact that two of them were fawns suggests that this was unlikely. Nevertheless, to be conservative, it seems certain that follicular activity becomes obvious at least thirty days in advance of breeding.

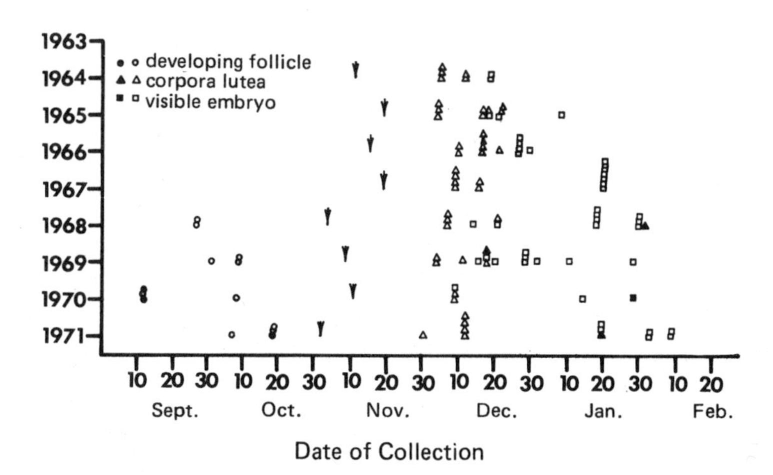

FIG. 4.2. Evidence of breeding activity in relation to date of collection for George Reserve female deer. Arrows indicate mean conception date of yearling and older females. Open symbols are yearling and older females; solid symbols are fawn females.

Recognizable embryos become apparent at about thirty days of age. Prior to this time, the luteal bodies in the ovary are indicators of reproductive activity. Only typical corpora lutea of pregnancy (see Mansell, 1971) were included in corpora luteal counts, but other current structures were used to indicate reproductive activity. Since histological preparations were not made and gross examinations under a dissecting microscope are not absolute, the possibility of errors in identification of structures is acknowledged. Nevertheless, the size and characteristic appearance of corpora lutea of pregnancy are such that errors are probably minimal. Furthermore (as discussed later in this chapter) the correspondence between embryo counts and corpora luteal counts in later stages of pregnancy is reassuringly close.

Therefore, on the basis of ovarian analysis, reproductive activity can be determined at least thirty days prior to the appearance of visible embryos, and frequently earlier. Thus, for female fawns (assuming that embryos are visible at thirty days and that fawns breed thirty days later than older females) tracts collected on or after the mean breeding date of older animals should show signs of reproductive activity—if breeding is to occur in that year. Those not showing signs of activity by that time are unlikely to breed as fawns. Conversely, female fawns taken prior to the mean breeding date must show positive signs in order to reach any conclusion about reproductive activity; those that are negative may or may not breed.

Changes in size and weight of the uterus accompany the maturation of follicles and formation of corpora lutea. Tracts of female fawns showing no sign of reproductive activity averaged a 7-g wet weight, while yearling and older animals averaged 15 g. Tracts showing a single developing follicle had a mean weight of 25 g, while those with two maturing follicles had a mean weight of 37 g. Tracts with one corpus luteum weighed 33 g; those with two corpora lutea had a mean weight of 48 g. Thus, the weight of the tract more than doubles prior to ovulation and continues to increase in size and weight during early pregnancy, before visible embryos can be found. The greater weight of tracts with maturing follicles and corpora lutea gives further indication of the breeding condition.

Embryo Rates of Yearling and Older Females

Females reproduced over the life span corroborated by field observations of very old, marked does that continued to produce healthy offspring. Yearling females tended to produce single fawns more frequently than older females, but twins were common at low population sizes. Triplets were recorded only in females older than 1.5 years.

Yearling females have a lower embryo rate than older females. Since 1963, twelve pregnant yearlings were examined which carried eight singles and four twins for a 1.33 embryo rate. Eleven 2.5-year-olds had four singles, five twins, and two triplets for a 1.82 embryo rate. There was little difference between 2.5- and 3.5-year old or older females, since twenty of the latter examined carried eight singles, eighteen twins and nine triplets for a rate of 1.83. Unfortunately, the sample size of yearling females was so small that this age class could not be treated separately. Therefore, the following analysis is based upon the combined yearling and older females. For convenience, the combined yearling and older females will be referred to as Y+Ad females.

In table 4.1 the results of reproductive tract examination are given for females that were obviously pregnant. The ratios of singles to twins to triplets show obvious changes over time, even taking into account the random variation that was present. Table 4.2 presents the same results by time period for Y+Ad females only. The relationship of number of embryos per female shows an apparent inverse relationship with the size of the posthunt population. The only serious departure was for the earliest period (1933–34 to 1937–38), and this was attributable to a time-lag effect, as discussed elsewhere (chaps. 3 and 11). However, if these years are eliminated from consideration, time-lag effects are minimized and a far more powerful analysis can be made by grouping females according to the size of the posthunt population, irrespective of year.

The data in table 4.2 do not take into account those females which failed to breed. With low populations the number was trivial, but as populations increased it became progressively more important. In recent years, complete reproductive tract examination gave fairly complete results. For earlier years it was necessary to examine the records closely and include as nonbreeders only those females collected very late in the harvest period, when visible embryos were being reported in other females. There is a small chance that some of these females listed as not pregnant were very late breeders, but errors from this source are likely to be insignificant. Recall that fawn females are not being included for the moment, and that they are the age class which presents most of the difficulty with late breeding.

TABLE 4.1. Numbers and Sexes of Embryos from Obviously Pregnant White-tailed Deer on the George Reserve

Year	*Number of Embryos*			*Sex*	
	1	*2*	*3*	♂	♀
34–35	3	10	0		
35–36	1	0	0		
36–37	3	8	0	4	6
37–38	5	5	1	1	2
38–39	1	4	0		
39–40	3	3	0		
40–41	4	4	0		
41–44	—	—	—	—	—
44–45	2	3	0		
45–46	4	2	0		
46–47	—	—	—	—	—
47–48	5	4	0		
49–50	1	1	0	2	1
50–51	—	—	—	—	—
51–52	1	13	0	15	12
52–53	0	1	0	2	0
53–54	8	5	0		
54–55	0	3	0		
55–57	—	—	—	—	—
57–58	5	12	3		
58–63	—	—	—		
63–64	5	7	1		
64–65	0	6	0		
65–66	0	5	0	—	—
66–67	6	3	0		
67–68	10	6	0	4	5
68–69	2	6	1	6	7
69–70	4	3	2	9	0
70–71	2	2	0	3	0
71–72	2	2	1	2	6
72–73	0	5	0	1	1
73–74	6	5	1	7	7
Σ in category	83	128	10	56	47
Percent	37.6	57.9	4.5	54.37	45.63
Σ of number	83	256	30		
Percent	22.5	69.4	8.1		

TABLE 4.2. Variation of Embryo Rate of Obviously Pregnant Yearling and Older Females by Time Period

Years	$\bar{X}$ Posthunt Population	Number Embryos 1	2	3	Σ	Sample Size	Embryos Per Female
33–34 to 37–38	138	12	23	1	61	36	1.69
38–39 to 40–41	101	8	11	0	30	19	1.58
44–45 to 56–57	86	20	32	0	84	52	1.62
57–58 to 63–64	60	9	19	4	59	32	1.84
64–65 to 67–68	73	16	20	0	56	36	1.56
68–69 to 70–71	63	7	13	4	45	24	1.88

The embryo rate and effect of nonbreeding according to the posthunt population number is shown in table 4.3. The number of embryos per pregnant female increased rapidly as population size declined and was due to a higher proportion of twins and triplets. At higher populations, the lower rate of embryos per female was accompanied by an increase in the number of nonbreeding females. There was a strong correlation between the number of embryos per Y+Ad female and the percent of those females that failed to breed (fig. 4.3). Thus, as the embryo rate declined, the proportion of nonbreeding females increased. Conversely, as the embryo rate increased, the proportion of nonbreeding females decreased. At an average embryo rate of 1.7, virtually all Y+Ad females bred.

TABLE 4.3. Relationship of Yearling and Older Female Embryo Rates and Nonbreeding to Posthunt Population of the Same Females

Posthunt Number	Total Sample	Number Pregnant	Percent Pregnant	Embryos Per Pregnant Female	Embryos Per All Females
10–20 ($\bar{X}$=19.0)	62	61	98.4	1.74	1.71
21–30 ($\bar{X}$=26.2)	57	52	91.2	1.71	1.56
31–40 ($\bar{X}$=36.3)	63	50	79.4	1.54	1.22
41+ ($\bar{X}$=49.8)	51	36	70.6	1.69	1.20

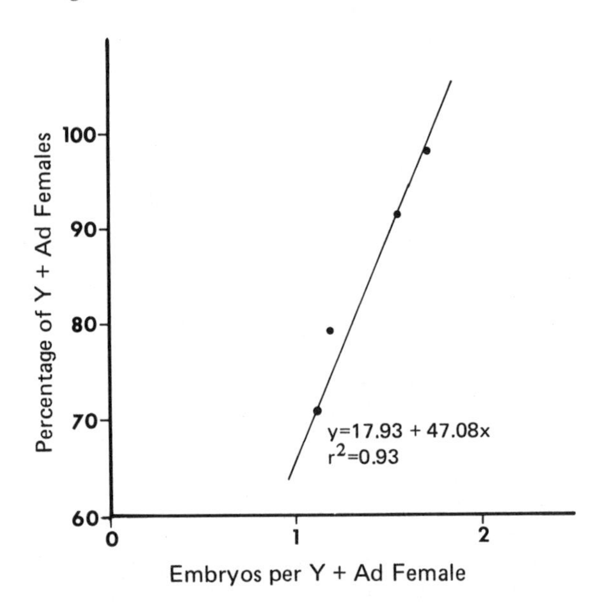

FIG. 4.3 Correlation of percentage of pregnant yearling and older females with embryos per yearling and older female. Data taken from table 4.3.

Embryo Rate of Fawn Females

It is apparent that at low population sizes fawn females represented a significant factor in reproduction, while at the higher populations they did not breed at all. As previously noted, even with the most careful examination of reproductive tracts it is difficult to get a reliable estimate of embryo rates because of the late date at which many fawns breed. Only in the last few years have such complete examinations been made; thus, the sample is small and represents only a small range of population sizes. Fawn reproduction rates must, therefore, be determined indirectly and then compared with the direct evidence available.

The regression of fawn recruitment on size of the Y+Ad female population is given in figure 6.1. It is a realized recruitment based on empirical data derived from the population and did not involve an estimation. This regression is shown in figure 4.4, along with a plotting of the embryo rate for Y+Ad females based upon both pregnant and nonpregnant individuals as given in table 4.3. The latter rate appears to show a linear relationship. Part of the

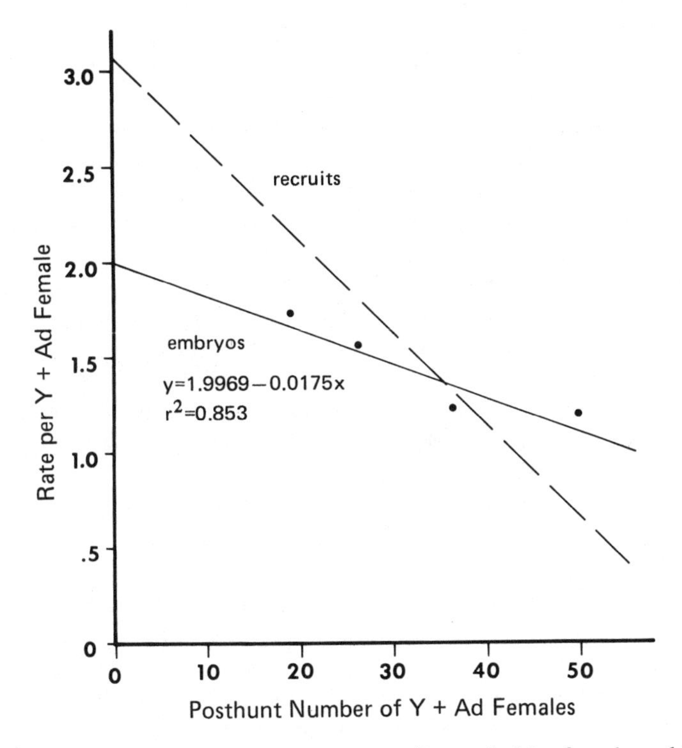

FIG. 4.4. Regressions of embryos per yearling and older female and fawn recruitment rate on the posthunt population of yearling and older females

deviation was due to the variable number of yearling females in the population, but sample sizes were too small to treat this age group separately.

The regression over the range the data covered is linear. Although the embryo rates depicted in the figure were obtained from combined data, treating the individual values in the regression gives virtually identical results, but with a lower coefficient of determination ($r^2 = 0.2236$). Thus, embryo rate showed a negative relationship to population size. As population size increased, embryo rates declined. This agrees with the results of embryo rates for white-tailed deer on the Seneca Army Depot in New York (Hesselton et al., 1965; Gross, 1969), and even the values seem similar if the fact that yearling and older females were combined in this study is taken into account. Teer et al. (1965) also reported that ovulation rate and the percentage of females that conceived were negatively related to population density.

The point at which the two regression lines of figure 4.4 cross is the point at which the young *in utero* of Y+Ad females equal the young recruited. That is, each embryo in a Y+Ad female in the winter was matched by one live fawn in the population the following fall. To the left of the crossing of the regression lines, the number of fawns recruited *exceeded* the number of embryos in Y+Ad females, which gave rise to the recruitment. The embryo rates at this end of the regression were the most reliable and involve almost no nonbreeding females (3 out of 108, or 2.8 percent were nonbreeding). This represents virtually the maximum reproductive effort that the Y+Ad females could have made. The excess between this maximum Y+Ad female effort and the realized recruitment must have been due to the breeding of fawn females.

Conversely, the difference between the regression lines to the right of their crossing represents embryos in Y+Ad females that were in excess of the fawns recruited in the fall. Therefore, the difference was the young that failed to survive sometime between the early embryonic stages and six months of age.

The rates shown in figure 4.4 are converted to numbers in figure 4.5. The peak in numbers of fawns recruited was achieved at a Y+Ad female population of 33, but this included a contribution from fawn females (10.2 percent). The maximum number of fawns produced by fawn females (9.1) occurred below the maximum recruitment point—in this case at a population of 20 Y+Ad females. It is apparent that the maximum possible reproduction, with all embryos surviving, was not achieved. On the basis of embryos, the maximum recruitment would have occurred at a Y+Ad population of 55 females if all embryos had survived. Even at the extreme right end of the recruitment curve, the potential in the form of embryos was still hardly diminished for the population. However, on a per individual basis, the embryo rate was declining (fig. 4.4).

The number of fawns recruited by fawn females can be determined directly from figure 4.5, but the rate per individual fawn female must be estimated from the approximate number of fawn females in the population. The regression of posthunt fawn females on the number of all posthunt females is given in figure 4.6. Although the regression was significant, the variation is readily apparent and was due to (1) an initial variation in the sex ratio at recruitment, plus (2) a variation in the number of female fawns in the hunting kill. These were observed data, and even though the variation between years was great, the mean values represent the best estimate over a series of years.

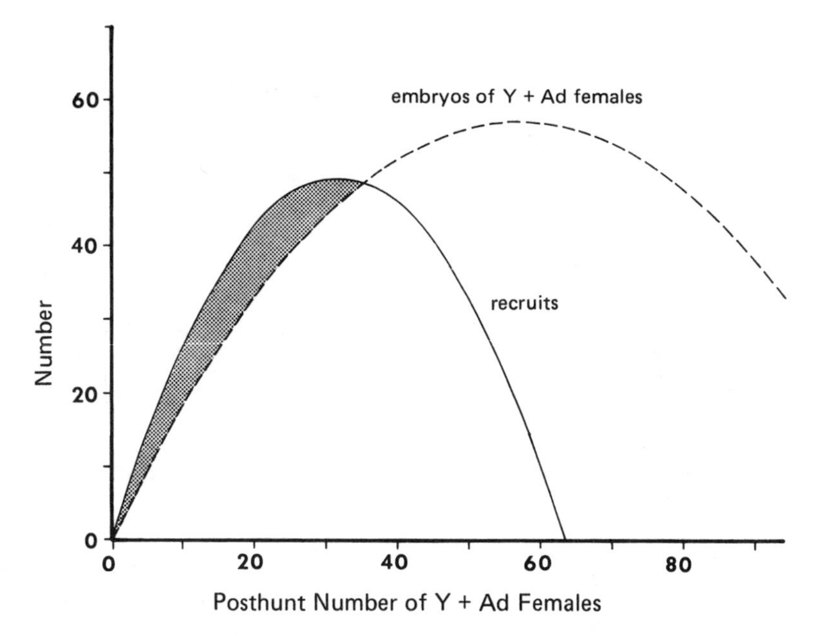

FIG. 4.5. Relationship of number of embryos and number of fawn recruits to posthunt population of yearling and older females. Stippled area indicates number of fawn recruits which had to be produced by fawn females.

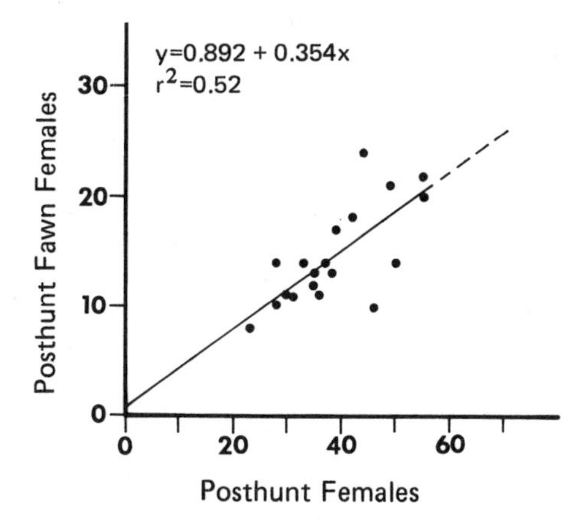

FIG. 4.6. Regression of posthunt fawn females on all posthunt females (including fawn females) for nineteen years

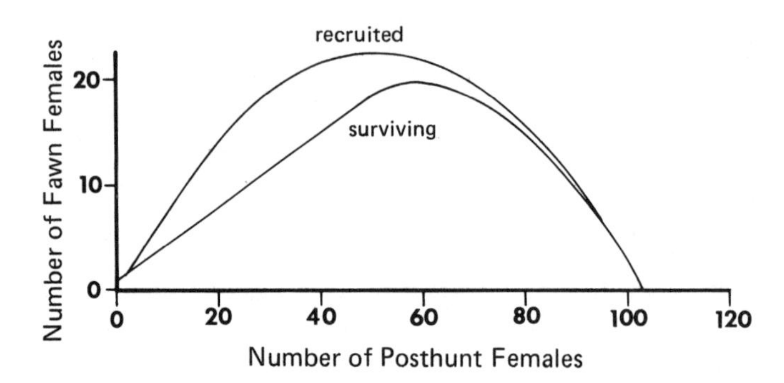

FIG. 4.7. Relationship of female recruits and posthunt survivorship of fawn females to the population of posthunt females of all ages. Data are taken from table 8.1.

The regression depicted in figure 4.6 was used to predict the number of fawn females over the range of posthunt total females (from 0 to 55). Projection of the regression to the left presents little problem since it must pass through the zero point; the regression equation, which was not forced through zero, nearly does so without this restriction. The relationships are shown in figure 4.7, where the number of recruits and posthunt survivorship of fawn females to the posthunt population are plotted on the posthunt number of all females. The depression of posthunt fawn females below the recruitment level (left-hand part of the graph) was due to hunting removal. On the right, continued low hunting kills and the invulnerability of female fawns resulted in higher survivorships.

Summarization of the analysis is presented in table 4.4 and figure 4.8, where the breakdown of embryos produced by Y+Ad females and fawn females is given. The embryo rate for Y+Ad females increased with declining population size to a maximum of approximately 2.0; for every Y+Ad female which had a single fawn, another must have had triplets. The embryo rate for female fawns increased to a maximum of approximately 1.62. (One case of twin embryos in a fawn female was recorded after the period of this study.) Fawn females declined in breeding as population size increased and dropped to zero when Y+Ad females averaged 1.4 fawns recruited per female. At low populations, with the combination of high breeding rates and prevalence in the population, fawn females accounted for a high percentage of the recruitment (e.g., at a population size of ten females, 47.7 percent).

TABLE 4.4. Relationship of Total Posthunt (PoH) Population (N) Size, Number of Female Yearlings and Older Females, and Yearling Fawn Females and Their Respective Embryo Rates.

Total PoHN	*Total* PoH♀♀	*Embryos/* Σ♀♀	PoHN Y+Ad ♀♀	*Embryos/* Y+Ad ♀	PoHN *Fawn* ♀♀	*Embryos/* *Fawn* ♀
10	5.23	1.78	2.49	1.95	2.74	1.62
20	10.45	1.70	5.86	1.89	4.59	1.45
30	15.68	1.61	9.23	1.84	6.45	1.28
40	20.91	1.51	12.61	1.78	8.30	1.11
50	26.13	1.41	15.98	1.72	10.15	0.94
60	31.36	1.32	19.36	1.66	12.00	0.77
70	36.59	1.22	22.73	1.60	13.86	0.60
80	41.82	1.12	26.11	1.54	15.71	0.43
90	47.04	1.02	29.48	1.48	17.56	0.26
100	51.75	0.93	32.97	1.42	19.30	0.10
110	52.27	0.88	37.83	1.33	19.67	0.00
120	62.72	0.85	43.21	1.24	19.51	
130	67.95	0.82	49.13	1.14	18.82	
140	73.18	0.78	55.59	1.02	17.59	
150	78.41	0.72	62.57	0.90	15.84	
160	83.63	0.65	70.08	0.77	13.55	
170	88.86	0.55	78.13	0.63	10.73	
180	94.09	0.44	86.72	0.48	7.37	
190	99.31	0.31	95.82	0.32	3.49	
200	104.54	0.17	104.54	0.17	0.00	
210	109.77	0.08	109.77	0.08		
220	115.00	0.00	115.00	0.00		

Although the embryo rate data are incomplete, it is desirable to determine to what extent they are consistent with the results derived above. The breakdown of reproductive tract examinations by age are shown in tables 4.5 and 4.6. The reasons for presenting two tables instead of one, combined table are because the tables represent data collected during two levels in population density and because the examinations were less complete in the earlier years than in the latter. For the period included in table 4.5, the mean total female prehunt population was 69.0, while that of the Y+Ad female numbers was 47.4. The mean posthunt total female population was 47.6, while the Y+Ad female mean was 32.0. For the period included in table 4.6, the mean female prehunt population was 53.0 while Y+Ad female numbers averaged 33.0; posthunt mean sizes were 20.0 and 13.0, respectively. Note that the data for 1971–72 are not complete (not included in the nineteen-year totals

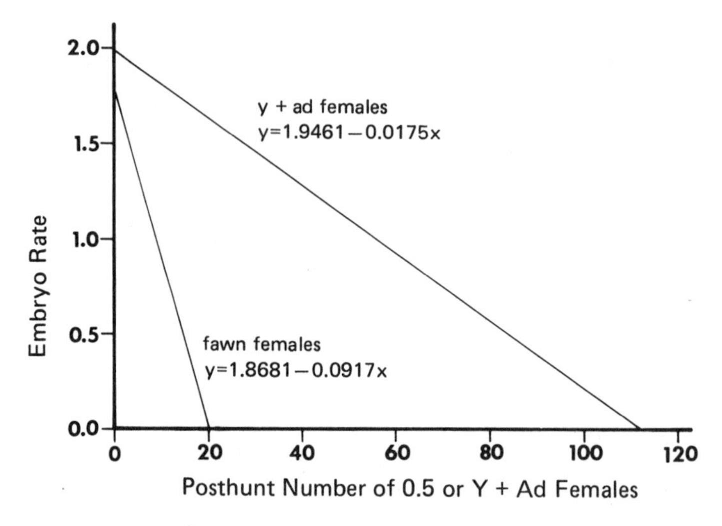

FIG. 4.8. Regressions of embryo rates on number of females in the yearling and older age category or fawn category

of the reconstructed population), but the possible errors from this source are minimal. Table 4.5 includes only mature (i.e., near ovulation) Graafian follicles, while table 4.6 includes all obviously developing follicles. In table 4.5, twenty-seven females had forty-four visible embryos for a rate of 1.63 embryos per female. In table 4.6, fifteen Y+Ad females had twenty-eight embryos for a rate of 1.87 per female. These results are comparable with the previous data presented on Y+Ad females because they are simply two ways of presenting the same information.

Interpretation of corpora lutea counts was somewhat difficult. Many authors (Cheatum and Severinghaus, 1950; Teer et al., 1965; etc.) have noted that corpora lutea counts usually exceed embryo counts. On the basis of females in this study which had obvious embryos, there were seventy embryos and seventy-four corpora lutea. Three females had more embryos than corpora lutea and seven had less, in all cases a difference of one. These were females from low population sizes, so the close correspondence of embryos to corpora lutea would be more likely than in higher population densities. Similar results were obtained for Iowa white-tailed deer by Haugen (1975). For the purposes of tables 4.5 and 4.6, if one equated corpora lutea to embryos, the result would be a very slight overestimation of the number of embryos. Twenty-nine Y+Ad females in table 4.5 had forty-six corpora lutea for a rate of 1.59 per

TABLE 4.5. Female Reproductive Tracts Examined for Embryos, Corpora Lutea, and Mature Follicles (Collected 1964–69)

Age	Number Examined	Number with Embryos 1	2	3	Σ of Pregnant ♀	Number with Corpora Lutea 1	2	Σ of Number with Corpora Lutea	Number with Mature Follicles 1	2	Σ with Follicles	Number Evidence of Breeding*
0.5	9	0	0	0	0	1	1	2	0	0	0	7 / 1
1.5	14	3	1	0	4	3	4	7	0	1	1	2 / 0
2.5	19	3	4	1	8	4	5	9	0	1	1	1 / 0
3.5	13	2	3	0	5	2	6	8	0	0	0	0 / 0
4.5	7	1	5	0	6	1	0	1	0	0	0	0 / 0
5.5	1	0	0	0	0	1	0	1	0	0	0	0 / 0
6.5	1	1	0	0	1	0	0	0	0	0	0	0 / 0
7.5	3	0	2	0	2	1	0	1	0	0	0	0 / 0
8.5	1	0	0	0	0	0	1	1	0	0	0	0 / 0
9.5	1	0	0	0	0	0	1	1	0	0	0	0 / 0
10.5	0	0	0	0	0	0	0	0	0	0	0	0 / 0
11.5	0	0	0	0	0	0	0	0	0	0	0	0 / 0
12.5	1	1	0	0	1	0	0	0	0	0	0	0 / 0
Total	70	11	15	1	27	13	18	31	0	2	2	10 / 1

*Number with no embryos or corpora lutea or follicles per number killed late enough to show breeding activity if present.

TABLE 4.6. Female Reproductive Tracts Examined for Embryos, Corpora Lutea, and Developing Follicles Collected from 1969–72.

Age	*Number Examined*	*Number with Embryos*			*Σ of Pregnant ♀*	*Number with Corpora Lutea*		*Σ of Number with Corpora Lutea*	*Number with Developed Follicles*		*Σ with Follicles*	*Number Evidence of Breeding**
		1	2	3		1	2		1	2		
0.5	15	3	0	0	3	1	0	1	4	0	4	7 / 3
1.5	12	2	0	0	2	2	4	6	1	2	3	1 / 0
2.5	7	0	1	1	2	0	2	2	0	2	2	1 / 0
3.5	7	2	3	1	6	0	0	0	0	1	1	0 / 0
4.5	1	0	0	0	0	0	1	1	0	0	0	0 / 0
5.5	3	0	1	1	2	0	1	1	0	0	0	0 / 0
6.5	0	0	0	0	0	0	0	0	0	0	0	0 / 0
7.5	2	1	1	0	2	0	0	0	0	0	0	0 / 0
8.5	3	0	1	0	1	0	0	0	1	1	2	0 / 0
9.5	0	0	0	0	0	0	0	0	0	0	0	0 / 0
10.5	0	0	0	0	0	0	0	0	0	0	0	0 / 0
Total	50	8	7	3	18	3	8	11	6	6	12	9 / 3

*Number with no embryos or corpora lutea or follicles per number killed late enough to show breeding activity if present.

female, while ten Y+Ad females in table 4.6 had eighteen corpora lutea for 1.80 per female. These rates compare very closely to the embryo counts. Correspondence of follicles to embryos is entirely unknown, but developing follicles indicate a probability that breeding will occur. Of the animals included in the two tables, not a single Y+Ad female could be established as having failed to breed.

It is with the fawn age class that the greatest interest lies. Of nine female fawns examined in table 4.5, none contained visible embryos; however one had a corpus luteum and another had two. Thus, at least two of the nine (22.2 percent) were reproductively active, and possibly one might have conceived twins. Of the remaining seven, only one was taken late enough in the season to be sure that breeding would not occur. A moderate amount of breeding activity by fawn females was apparent at this population size.

In table 4.6, of fifteen female fawns, three had visible embryos, one had a corpus luteum, and four had developing follicles. Of the seven not showing reproductive activity, only three were taken late enough to be sure that they would not breed. Thus, the breeding status of eleven female fawns could be determined and four were uncertain. Eight of the eleven (72.7 percent) showed pregnancy or signs of expected pregnancy. Assuming a single embryo, the rate would be 0.73 embryos per female.

Since the data of table 4.5 were from a mean population of 15.6 fawn females, and table 4.6 from a mean population of 11.0, it is concluded that the female fawn reproductive rates presented in table 4.4 are realistic. If it is assumed that evidence of breeding would result in a single fawn, the rates derived from the examination of reproductive tracts are 0.22 and 0.70, as compared with an expected 0.44 and 0.80 (table 4.7). Sampling errors due to small sample size, the failure of corpora lutea or follicles to be representative of actual embryos, and mortality *in utero* or after birth could easily account for these small differences. Also, the time of collection would tend to bias the data for Y+Ad females (which breed early) on the high side, while for the fawn females (which breed late) it would tend to bias the data on the low side. However, the agreement is quite close and it can be concluded that the results from reproductive tract analysis agree reasonably well with the recruitment results, and that the relationships presented in figure 4.8 and table 4.4 are essentially correct. Error, if any, would be due to a slight underestimation of embryo rates for Y+Ad females and a slight overestimation of those for fawn females (table 4.7).

The embryo rates of fawn females may represent the maxi-

TABLE 4.7. Expected Embryo Rates (from Regression in Figure 4.8) Compared with Observed Rates from Examination of Female Reproductive Tracts

	Using Total ♀♀ Regressions				
		Expected embryo rate		*Observed embryo rate*	
Period	*Total Number* ♀♀	Y+Ad ♀♀	Fawn ♀♀	Y+Ad ♀♀	Fawn ♀♀
64–65 to 68–69	47.6	1.43	0.23	1.63	0.22
69–70 to 71–72	30.5	1.79	0.84	1.87	0.73

	Using 0.5 or 1.5+ ♀♀ Regressions					
	Number ♀♀		*Expected embryo rate*		*Observed embryo rate*	
Period	Y+Ad	Fawn	Y+Ad ♀♀	Fawn ♀♀	Y+Ad ♀♀	Fawn ♀♀
64–65 to 68–69	32.0	15.6	1.39	0.44	1.63	0.22
69–70 to 71–72	19.5	11.0	1.72	0.86	1.87	0.73

mum achievable average under the conditions of soils, climate, and vegetation in the southern Michigan deer range. On the basis of 750 fawn females from southern Michigan examined by the Michigan Department of Natural Resources biologists between 1955 and 1975, a weighted embryo rate of 0.65 was obtained (data from Youatt et al., 1975). On the basis of small yearly samples, the range was from 1.20 to 0.43. Twins occurred fairly regularly. These results, obtained at a time when the deer density in southern Michigan was quite low, agree closely with the maximum average attained by fawn females in the George Reserve. For the same period, 1,649 Y+Ad females in the Department of Natural Resources sample had a weighed embryo rate of 1.60. If one regresses the embryo rate of fawn females on Y+Ad females for the George Reserve, these values from southern Michigan fall directly on the line. This would match the rate of George Reserve older females at a Y+Ad female density of about thirty-five (or about seventeen per square mile). Again, the correspondence appears to be quite good.

In the corn belt of Iowa, fawn females have been reported with embryo rates of 1.25 (Haugen, 1975), and both twins and triplets have been recorded. This variation in maximum rate by region could be accounted for easily by phenotypic response due to quality of diet in various environments. It is of further interest that Michigan Department of Natural Resources biologists working in the Upper Peninsula (U.P.) of Michigan have not been able to induce fawn females to breed by feeding standard rations *ad libi-*

dum; nor do female fawns breed on a mix of natural and artificial foods in a square mile enclosure at the Cusino Research Station in the U.P. (Verme, personal communication, 1977). This has led to speculation about possible genetic differences in northern deer. However, the George Reserve deer stock was obtained from Grand Island in the U.P., and this would argue against a genetic explanation. The difference would seem to lie with relative differences in the available nutrients of the two areas, or other environmental factors, such as temperature, daylength, etc.

It is significant that George Reserve fawns began to breed at a population density at which virtually 100 percent of the fawns survived. The correspondence of these two points was demonstrated both by recruitment data and the population density at which the appearance of embryos in female fawn reproductive tracts began. Thus, in sharp contrast to older females, fawn females only bred under circumstances in which the survivorship of the offspring was extremely high. It is further worth noting that the production of triplets by adult females and the breeding by fawn females appear to occur at approximately the same densities. This may be a useful clue to the breeding of fawn females in a wild population if further study affirms the relationship.

Sex Ratios

The sex ratio, based upon 110 embryos large enough to determine sex accurately (approximately 50 mm forehead-rump length), was 1.19 males:1 female (54.37 percent male and 45.63 percent female; table 4.1). Samples for most years were small and quite variable. For example, in 1969–70, 9 males and no females were recorded. Thus, little reliance can be placed upon results by year. However, from the total sample size, it appears that a nearly balanced sex ratio with a slight preponderance of males was emerging. This would agree with the results on white-tailed deer from other areas (Severinghaus and Cheatum, 1956; McDowell, 1959; etc.).

The sex of thirty sets of twins was determined. In eleven sets, the twins were both male; in ten sets, both female; and in nine sets, a male and a female. Thus, there does not seem to be a significant deviation from a random assortment of combinations. Two sets of triplets which could be sexed showed two males and a female, and two females and a male.

When the sex ratio of embryos was compared to the sex ratio of fawns recruited, close agreement was achieved (1.19:1, 1.15:1; 54.37 and 53.41 percent males, respectively). The recruitment sex ratio was based upon 822 fawns spanning nineteen years. Sex ratio at recruitment was the outcome of the primary sex ratio and the subsequent, early mortality of fawns. The close comparison of the sex ratio of embryos and that of recruits suggests that differential mortality by sex of young fawns was minor. In fact, differential mortality by sex did not occur until the deer reached the yearling age (see fig. 5.1). Early fawn mortality at the higher population densities was present, while at the lower densities it was negligible. Recruitment figures were essentially a total count, rather than a sample, and a good deal of reliance can be placed upon the results.

Verme (1965) found from penned feeding experiments that females on adequate diets produced predominantly female offspring, while those on restricted diets produced predominantly male offspring. Verme noted that the statistical significance (0.95 level) of his results could be questioned. However, he contends (and I concur) that the results strongly suggest that sex ratio was related to food availability and cannot, therefore, be passed off as a mere sampling error.

Most females on the George Reserve breed prior to the period of harvest, and if sex was related to population density, the prehunt population would be the appropriate value to use. The regression of percentage of male offspring on prehunt population showed a surprisingly good fit (significant at the 0.94 level) with the percentage of male fawns increasing as population density increased (figure 4.9). There was considerable variation between years, but in general there was a positive relationship between prehunt population size and percentage of male offspring. Based upon all years, this slope narrowly missed being significantly different from zero at

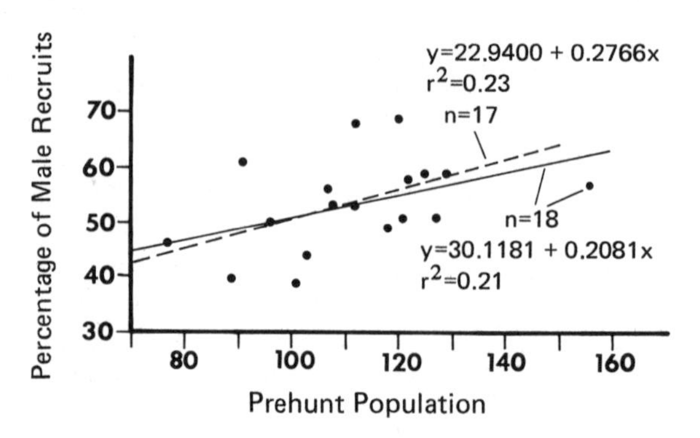

FIG. 4.9. Correlation of sex composition of offspring with prehunt population size

the 95 percent confidence level ($p = 0.06$). If the extreme right hand point on the figure is deleted upon the grounds that the regression may not be linear over the range of values (there may be an upper limit to the amount of imbalance that sex ratio undergoes), the slope is significantly different from zero.

Noting a percentage point in the statistical test may be begging the point. The trend is clear, and the relationship cannot be ignored. Furthermore, the results cannot be accounted for by differential mortality by sex since at lower population densities the number of embryos and the number of fawns recruited were the same, as were the number of corpora lutea and number of embryos. Hence, because there was virtually no mortality of offspring between ovulation and six months of age (recruitment) the sex ratio was not altered by differential mortality. These results confirm the observations of Verme (1965): at low populations there was a disproportionate number of female offspring produced, while at high populations there was a disproportionate number of male offspring.

An additional question concerns whether or not the change in sex ratio of offspring was related to sex distribution in the breeding population. Up to this point, the total population has been considered, and one could hypothesize that the numbers of one sex or the other could be the important variable. For example, suppose that the physiological mechanism responsible for the differential survival of x-bearing and y-bearing gametes during gametogenesis in the male and the environmental stimulus controlling the survival of x- and y-bearing gametes were both due to the prevalence of males in the population. Conversely, suppose that the physiological mechanism involved differential activity of x- and y-bearing sperm in the vagina and uterus of the female, due to physical and chemical characteristics influenced by the prevalence of females in the population. In these cases, one sex or the other could be controlling the sex ratio of offspring.

Least squares regressions of sex ratio of offspring were calculated for the population size of each sex separately. Neither fit was as good as with the total population (total population $r^2 = 0.2105$; for males only, $r^2 = 0.1675$; for females only, $r^2 = 0.1239$). Neither slope was significantly different from zero. Furthermore, regression of sex ratio of offspring against sex ratio of parents gave a random distribution of points with essentially zero slope ($r^2 = 0.0001$; $b = -0.0116$). It was concluded, therefore, that the varying sex ratio of offspring was related to the total population, irrespective of sex. (Further discussion of sex ratio selection can be found in chapter 14.)

CHAPTER 5

Mortality

The schedules of survivorship for sex and age will be considered in this chapter. The purpose is to determine the probability of further life (or conversely, death) for given sex and age classes.

Analysis of sex composition of the kill is possible over the forty-two years for which kill records have been kept (1933–34 through 1974–75). However, age data are mainly lacking for the first nineteen years, except for the zero age class (fawns); and even in that category some errors are likely. Therefore, most of the analysis of age at mortality is based upon the nineteen years for which reconstructed population data are available.

During the period 1952–53 through 1974–75, when all lower jaws were saved for age determination, shooting by man was virtually the sole cause of mortality of recruitment age and older deer. No known case of mortality was due to causes usually referred to as "natural" (i.e., starvation, predation, disease, etc.). A few accidental deaths have occurred. Death of fawns from broken necks has occurred when they panicked and leapt against the fence during drive census or in front of the drive line when deer were about to be harvested. A few other deer have been sacrificed when they suffered serious injuries during trapping and marking. But all other cases of death have been directly attributable to serious gunshot wounds.

Two major factors are involved in the probability of death: vulnerability due to behavior of the deer, and selectivity by shooters for given sex- or age-classes. The following analysis will attempt to separate the two.

Differential Mortality by Sex and Age

Males were more vulnerable to mortality than females (tables 3.1 and 3.2). The sex ratio at birth was unbalanced in favor of males,

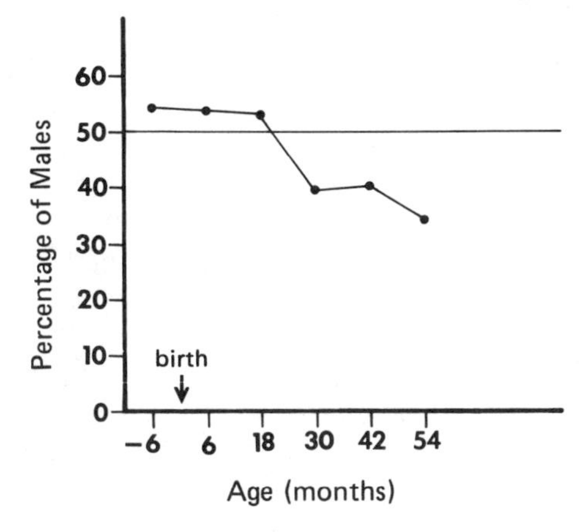

FIG. 5.1. Shifts in sex composition with age

with embryo counts showing 54.37 percent males (table 4.1). At recruitment age (± 6 months) there were 53.41 percent males. This unbalance occurred to the beginning of the harvest of the yearling year, at which time yearlings were 52.93 percent males. Thereafter, a disproportionate number of the killed animals were males (54.57 percent; fig. 5.1).

For all age classes combined, females were in a slight preponderance in the prehunt population (52.73 percent females; tables 3.1 and 3.2), and because of a disproportionate kill of males, the posthunt population was further unbalanced towards females (57.48 percent), as shown indirectly in figure 5.1.

More direct comparisons are made in table 5.1 for various time

TABLE 5.1. Comparison of Sex Ratio in the Population with Sex Ratio in the Kill by Time Periods for Various Harvest Methods

Methods	*Percent in Population*			*Percent in Kill*		
	N	♂	♀	N	♂	♀
Caretaker removals						
1933–34 through 1952–53	—	—	—	767	53.59	46.41
Shooters on stand						
1953–54 through 1967–68	1700	47.41	52.59	663	55.51	44.49
Kill from vehicle						
1968–69 through 1974–75	476	43.91*	56.09*	278	47.12	52.88

*Reconstructed population not complete for 1971–72 onward. Sex ratio based upon data available.

periods when different harvest methods were used. The first kills were made in 1933–34, and until 1952–53, most of the deer were killed by the first reserve caretaker, Larry Camburn, and a few other reserve employees. The sex composition in the population during that period is not known. The period when deer were driven past shooters by a line of beaters includes 1953–54 through 1968–69. The actual sex ratio in the population during this period can be determined from the reconstructed herd data. The last period involved shooting by me or the current caretaker, Dick Wiltse. Most of these animals were taken by spotlight, at dawn or dusk; the shooting was done from a vehicle. The reconstructed herd data are not complete for the entire period, but the sex ratio for the kill is known.

It is clear from table 5.1 that the kill in the years of shooters on stand resulted in more males than females being harvested. It is very likely true for the early years, as well, when the caretaker did the harvesting. In one year (1955–56) an attempt was made to kill only males, and thirty-six of the forty animals killed were males. Nevertheless, this year alone does not begin to account for the total discrepancy. Most of the disproportionate kill of males can be attributed to selectivity by shooters. A number of shooters on stand were competitive and, given the opportunity, selectively shot males.

By contrast, shooting from a vehicle in recent years was strongly geared to efficiency, and the first clear target was accepted without respect to sex or age. The discrepancy of the kill composition from the actual was much less. Some deviation was still present, but it can be accounted for by the differential vulnerability of some males, as will be discussed later in this chapter.

It is concluded that until very recent years a disproportionate number of males were killed from the population, and a good part of the difference was due to hunter selectivity. The effect on the population was to shift the composition from a disproportionate number of males at birth to a disproportionate number of females in the total population. Thus, the mean life expectancy of a male at birth was substantially less than that of a female. Note, also, that relationships found for legal harvest within the reserve closely matched the results from poaching (chap. 2).

Differential mortality by age can be examined by comparing the proportions of age-classes in the kill and prehunt population, as obtained from the reconstructed herd (tables 3.1 and 3.2). If pronounced differences in these proportions exist, then either vulnerability or hunter selectivity or both varies with age.

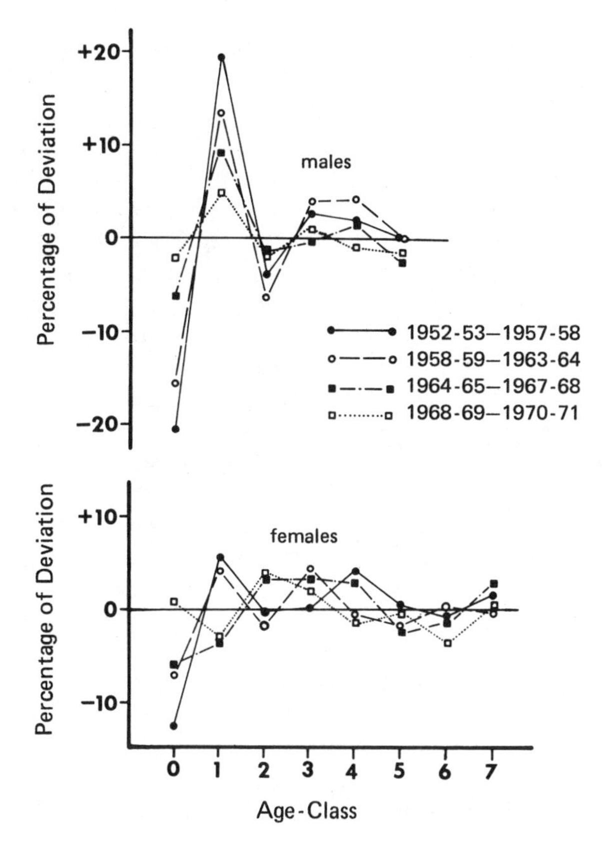

FIG. 5.2. Comparison of deviations of the proportion of age-classes in the kill from those in the prehunt population. Male animals 5.5 years and older and females 7.5 years and older were combined.

These comparisons are shown in figure 5.2 for both sexes. Various hunting methods are separated to show the shifts in hunter selectivity over time. The kill of the early drive method with shooters on stand showed greatest hunter selectivity. Hunter selectivity declined in later years, apparently because greater emphasis was placed on the shooters being nonselective. Hunting by spotlight, the most recent practice, was least selective.

For males, the pattern was consistent over the years. Male fawns were consistently underrepresented in the kill. This remained true even under the least selective hunting practices, as carried on since 1968–69. Yearling males, on the other hand, were highly vulnerable to shooting, regardless of shooting selectivity. The similar vulnerability of yearlings was reported for white-tailed deer in New

York by Maguire and Severinghaus (1954) and in Illinois by Roseberry and Klimstra (1974), and for black-tailed deer in California by Taber and Dasmann (1954). Males 2.5 years old were consistently underrepresented in the kill. Some variation occurs in older age-classes, but it was greater during the earlier periods and may have been due largely to hunter selectivity. Three and one-half- and 4.5-year-old males were somewhat overrepresented in the kill, while 5.5-year-old and older males were slightly underrepresented.

These results would seem to be expected in view of the biology of males. The survivorship of male fawns is very high (nearly equal to females) while still in the company of the female (fig. 5.1). With the birth of the next set of offspring, the mother drives her yearling male offspring away, and they establish new social relationships. It is during this period that the inexperienced young males are highly vulnerable to hunting mortality. However, those males that survive the yearling stage develop experience, and their survivorship improves. Although sexually mature, they are so low in the dominance hierarchy that they seldom engage in overt sexual behavior and are usually relegated to bachelor groups of young males. The ages 3.5 and 4.5 are ones of greatly increased vulnerability, perhaps because of increased exposure during rut as the young males move higher in the hierarchy. Older males, due to their greater experience, are able to minimize hazards associated with rutting behavior.

Hunter selectivity magnified the biological vulnerabilities and accounted for about two-thirds of the variation of mortality of males by age class (fig. 5.2). Still, there was a residual effect that was present even under the least selective hunting (1968–69 through 1970–71). Thus, it seems that behavior of the male influences the rate at which various age-classes are placed in situations of danger.

The pattern of deviations for females was neither so pronounced nor so clear as for males, and deviations by hunting method were less consistent. In general, fawns seemed to be lower in vulnerability than older females, and females aged 1.5 years through 4.5 years were more vulnerable. Vulnerability of females at ages 5.5 and 6.5 seems to be lower than expected, while that of 7.5-year-old and older females increased. This general pattern seemed so consistent that it was unlikely to be random variation.

Hunter selectivity was probably responsible for the low proportion of female fawns in the kill. Discrimination against the shooting of fawns of both sexes by sport hunters was reported by Maguire and Severinghaus (1954) and Roseberry and Klimstra

(1974). Note that the proportion of female fawns killed by the least selective hunting method was approximately the same as in the population. Since age is poorly correlated with size in the mature females, hunters could be selective on the basis of size but not age after the first two years. In fact, many yearling females approach adult size, and it is questionable whether they could be distinguished under field conditions. Thus, most of the differences in figure 5.2 for Y+Ad females could be attributed to vulnerability due to behavior of the animals.

Greater experience and improved hierarchial position could account for the decreased mortality rate of older females, but why the oldest females became more vulnerable was not clear. Perhaps they began to lose vigor and declined in hierarchial position or became more vulnerable because of the need to move about more to obtain equal quality resources. Such suggestions are highly speculative, and the answer must await detailed behavioral work on females of known age and identity.

These results were obtained from harvests over most of the fall and early winter. The reader should be cautioned that different hunting seasons might produce somewhat different results. Vulnerability between sexes and ages may be strongly influenced by the timing of the hunting season in relation to the rutting period.

Influence of Parental Care on Survival of Offspring

There are two aspects of parental care which may be related to the survivorship of offspring in the deer population. First is the cost of the young to the mother in the parental investment sense of Trivers (1972). Males do not contribute to parental care. The cost to females would be highest in the nursing state but would drop to a very low (perhaps insignificant) level following weaning. Mothers no longer assume obvious risk in the face of predators or directly protect the offspring from domination by and competition from conspecifics. However, some benefit may be indirectly derived by the offspring remaining close to a dominant mother. Benefit to the young consists primarily of being able to follow an experienced animal. Cost to the parent in being followed is probably slight, and, in fact, antipredator benefits may be substantial (Hirth and McCullough, 1977).

The second aspect of parental care affecting offspring survival involves termination of all benefits to the young, including leadership. In the strict sense, termination of parental care for the male offspring does not ordinarily occur until one year of age, when typically the female drives off her yearling offspring prior to the birth of new young. The male seldom rejoins the family group and usually joins another bachelor male group at this time (Hirth, 1977). The female yearling often is allowed to rejoin the mother when the new offspring are several weeks old, and she is able to follow at heel. Thus, for females, parental care typically lasts two years, and perhaps longer.

The relative importance of parental investment versus total parental care can be evaluated in terms of sex ratio changes with age on the combined data of embryos and recruits (fig. 5.1). It can be seen that the sex ratio through eighteen months of age differed only slightly from that of the embryo state. Males constituted 54.37 percent of 103 embryos, 53.41 percent of 822 six-month-old fawns, and 52.93 percent of 580 eighteen-month-old deer. Clearly, the sex ratio did not become balanced until after eighteen months of age, when males died at a much higher rate than females. These results suggest that it was total parental care rather than parental investment that was the important factor. Female yearlings continued to derive benefits from the mother in the form of experienced leadership, of which male yearlings were deprived. These results will be discussed further in chapter 14.

Survivorship Schedules

Population analysis by life table methods has been widely used in ecology since its value was established for human mortality schedules by actuaries (Deevey, 1947; Caughley, 1966; Mertz, 1970; and many others). In theory, life table methods should be a powerful analytical tool, particularly with populations of K-selected organisms, such as deer. Recently, a number of workers have begun to question the practical use of life tables, most notably Caughley (1966). Caughley reviewed a number of published life tables and concluded that most were invalid due to inadequate sample size, failure to meet the assumption of a stable age distribution, strong bias in sampling of the zero age-class, and confusion as to whether the sample represented animals living or having died.

To these functional problems should be added the philosophical question of what it is that a life table demonstrates. There has been a tendency to view schedules of survivorship and fecundity as ''typical'' or ''normal'' for a species, when they actually represent population response to a given set of environmental circumstances. Since environments vary over time and space, a given life table represents only one of almost an infinite number of possible states. The applicability of a given life table to other areas or other times is entirely dependent upon the homogeneity of the environment over time and space. Such conditions may hold in some circumstances, but variation is the more common case. The ''balance'' of nature is a dynamic one, involving many processes and characterized by time lags—a far cry from perfect tracking of cause and effect. It is probably useful to remind ourselves occasionally that the mean with a low variance that we strive so mightily to produce is an artifact. Even in protected areas there are good and bad years, fluctuations of competitors and predators, and numerous other factors that can influence the survivorship and fecundity of a species. Certainly periodic cycles, some of the most spectacular fluctuations to be found in nature, were manifest in the absence of modern technological man.

The George Reserve, being essentially a natural area, would be expected to approximate a steady state. Furthermore, attempts at managing the harvest of deer have been geared towards increasing the stability. Granted, natural predators would function in a different (i.e., selective) manner, but there is reason to suppose that fluctuation might be even greater with natural predators, as will be discussed later. It suffices for the moment to assume that controlled human hunting in the absence of effective natural predators would increase stability.

If a steady state exists, one would expect that life table analysis of the data on the George Reserve deer herd would be a population ecologist's dream. Consider that the data for each year presented life table series for vertical (time-specific) tables—nineteen of them for each sex (tables 3.1 and 3.2). Furthermore, the diagonal series for year-classes (beginning with a year-class and moving one row down for each column across to the right) gives at least seven complete horizontal (dynamic) life tables for each sex.

Nevertheless, the life table analysis is a disappointment. The first problem encountered is the small sample size, even though in fact it is the total population. An unanticipated difficulty arose from having certain age classes unrepresented (e.g., in 1952–53 there were no five- through eight-year-old males, but there were nine- and eleven-year-old males) since life table calculations cannot handle zeros in a l_x series.

TABLE 5.2. Comparison of Survivorship on the George Reserve for the Combined Data for Nineteen Years Calculated Directly from the Prehunt Population and Indirectly from the Kill

	♂♂		♀♀	
x	*Prehunt* l_x	*Kill* l_x	*Prehunt* l_x	*Kill* l_x
0	1.000	1.000	1.000	1.000
1	.699	.692	.713	.720
2	.275	.251	.460	.460
3	.175	.176	.298	.291
4	.075	.075	.167	.164
5	.021	.022	.094	.093
6	.014	.015	.073	.066
7	.009	.011	.055	.050
8	.007	.007	.024	.021
9	.007	.007	.010	.008
10	.005	.004	.005	.005
11	.005	.002	.003	.003
12	.005	.002	.003	.003

The d_x series does not present this problem. Comparison of the life tables calculated from l_x and d_x series on the combined data gave good correspondence (table 5.2). The differences were not significant when tested by chi-square. Caughley (1966) has pointed out that removals give an estimate of the composition of the living population.

Do the data give a reasonable approximation of a stable age distribution over the nineteen years, or for some smaller sequence of years? Certainly there was not a stationary age distribution, in which a stable age distribution and an unchanged total population size was present. From figure 3.1 it is obvious that the population level fluctuated greatly over the nineteen years. Furthermore, there was no sequence of years in which an approximation of a level population was present over a period approaching the longevity of a year-class (thirteen years). The maximum time for which a reasonably similar total population existed was four years (1965–66 through 1968–69). Therefore, there is no way of comparing the seven years of dynamic life tables with a corresponding set of time-specific tables, since agreement would be achieved only if a stationary age distribution were present. However, there is still considerable interest in whether or not there is an approximation of a stable age distribution. The fact that the total population fluctu-

TABLE 5.3. Mean Percentage of Population by Age Class for Nineteen Years of Vertical Life Tables (l_x Series)

	Males			*Females*		
Age Class	*Mean %*	*SD*[1]	*Range*	*Mean %*	*SD*[1]	*Range*
0	43.95	7.27	27–54	34.79	6.28	25–45
1	30.63	6.08	21–41	24.37	6.28	14–41
2	11.58	4.65	6–22	15.84	5.28	6–25
3	7.79	4.37	2–19	10.26	4.01	5–18
4	3.21	2.35	0–7	5.47	3.31	0–11
5	0.79	1.47	0–5	3.32	2.43	0–7
6	0.58	1.07	0–3	2.58	2.04	0–7
7	0.37	0.90	0–3	1.84	1.80	0–6
8	0.21	0.63	0–2	0.79	1.36	0–4
9	0.37	0.90	0–3	0.26	0.65	0–2
10	0.11	0.46	0–2	0.11	0.32	0–1
11	0.16	0.50	0–2	0.05	0.23	0–1
12	0.05	0.23	0–1	0.05	0.23	0–1

[1]Standard deviation.

ated during the nineteen years means that the possibility of a stable age distribution is highly unlikely. This could happen only if the mortality was totally random by age-class and the recruitment rate remained constant. Figure 3.1 conclusively shows that this was not the case.

While we can be sure of the conclusion that a strict stable age distribution was not present, we cannot be sure how greatly the years (or year-classes) varied from each other without further analysis. This exercise would hardly be worthwhile in terms of what it tells us about the George Reserve deer herd. However, it is seldom that the data are available to perform such an analysis, and it is most instructive about the application of the method to unrestrained populations.

The variance of proportions in the age-classes over the nineteen-year period for vertical tables is shown in table 5.3; the data for ten years (the error of very old animals for the three additional years beyond the seven complete years to the twelfth age-class was considered trivial) of horizontal tables is shown in table 5.4. It can be seen that there is considerable variation.

Another way to examine the possibility of a stable age distribution is to compare the proportions of a given year (vertical table) with that of the following year by a chi-square goodness-of-fit test.

TABLE 5.4. Mean Percentage of Population by Age Class for Ten Years of Horizontal Life Tables (l_x Series)

	Males			*Females*		
Age Class	*Mean %*	*SD*[1]	*Range*	*Mean %*	*SD*[1]	*Range*
0	43.20	5.55	37–57	34.10	5.11	26–43
1	33.00	5.87	20–40	26.20	4.57	18–32
2	10.90	3.38	6–15	15.60	4.95	11–28
3	7.90	3.11	5–13	9.80	2.10	7–14
4	2.60	1.90	0–6	5.30	3.30	0–10
5	0.40	0.84	0–2	3.10	1.97	0–6
6	0.40	0.84	0–2	2.30	1.95	0–6
7	0.40	0.84	0–2	1.70	1.64	0–5
8	0.40	0.84	0–2	0.50	0.97	0–3
9	0.40	0.84	0–2	0.20	0.42	0–1
10	0.40	0.84	0–2	0.20	0.42	0–1
11	0.0	0.0	0–0	0.20	0.42	0–1
12	0.0	0.0	0–0	0.10	0.32	0–1

[1]Standard deviation.

This tests the null hypothesis that there is no difference in the distribution of the proportions by age-class, a situation which would hold if a stable age distribution were present. Rejection of the hypothesis leads to the conclusion that the age classes were not stable between the two years. A similar procedure can be used to test for differences between successive year classes (horizontal tables). The statistical results are given in table 5.5. In both vertical and horizontal tables, males more closely approximated a stable age distribution than females. Statistical differences between pairs of years for females occurred in eleven out of eighteen comparisons of vertical tables and four out of nine horizontal tables. Furthermore, the longest string of consecutive, stable tables (i.e., not significantly different) was three—not a very good string in view of the longevity of thirteen years. Males showed better, with thirteen out of eighteen vertical comparisons and seven out of nine horizontal evidencing no significant difference. Closer examination, however, shows the figures to be misleading. Although no significant difference may be present between consecutive years, a stable age distribution would require that none of the years be significant from any of the other years. Intercomparisons of the string of seven showed seven out of fifteen intercomparisons of nonconsecutive years to be significantly different. It is concluded, there-

TABLE 5.5. Summary Table of Statistical Tests of Differences of Proportions in Age Classes

Comparison		*Vertical Tables* ♂♂	♀♀	*Horizontal Tables* ♂♂	♀♀
(19)52–53	with (19)53–54	HSD	HSD	NSD	NSD
53–54	54–55	HSD	HSD	NSD	NSD
54–55	55–56	SD	HSD	NSD	HSD
55–56	56–57	NSD	HSD	NSD	NSD
56–57	57–58	NSD	SD	NSD	HSD
57–58	58–59	NSD	SD	NSD	HSD
58–59	59–60	NSD	NSD	SD	NSD
59–60	60–61	NSD	SD	HSD	NSD
60–61	61–62	HSD	NSD	NSD	HSD
61–62	62–63	NSD	HSD		
62–63	63–64	NSD	NSD		
63–64	64–65	NSD	NSD		
64–65	65–66	NSD	NSD		
65–66	66–67	NSD	SD		
66–67	67–68	NSD	HSD		
67–68	68–69	NSD	HSD		
68–69	69–70	SD	NSD		
69–70	70–71	NSD	NSD		
ΣHSD		3	7	1	4
Σ SD		2	4	1	0
ΣNSD		13	7	7	5

Note: No significant difference (NSD) indicates no significance at the 0.05 level; significant difference (SD) indicates significance at the 0.05 level; highly significant difference (HSD) indicates significance at the 0.01 level.

fore, that the assumption of a stable age distribution cannot be made on any of the possible life tables within the data. This agrees with the conclusion of Miller (1976) about long-lived species.

Although a stable age distribution was not present (due to great variation between years and a small sample size), let us assume that the population was precisely controlled at a given posthunt population by shooting for a series of years. At the end of the series, an average age structure could be determined for that population equilibrium. The necessary data are available to derive these average survivorships for the George Reserve deer population. The zero age class is based upon embryos, rather than number of young born. Recruitment gives survivorship to six months of age, and vulnerability to shooting can be based on the least selective shoot-

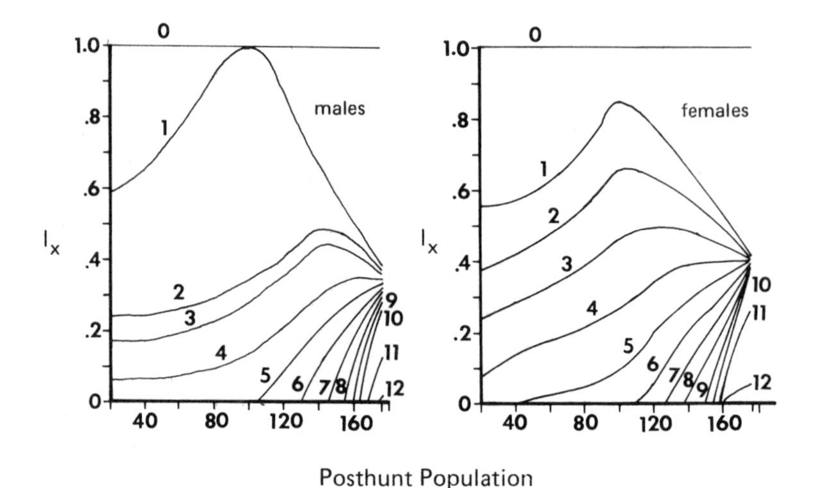

FIG. 5.3. Changes in average survivorship (l_x) of various age-classes (numbers on the curves) dependent upon population size assuming equilibrium

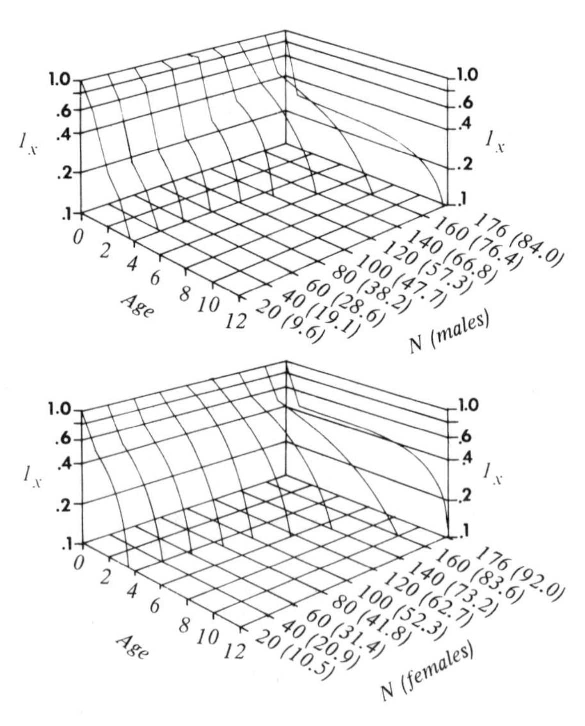

FIG. 5.4. Survivorship curves at equilibrium for various posthunt population sizes

ing (fig. 5.2). Females were considered essentially equally vulnerable, because kill by age class was similar to their proportion in the population. Male kills were apportioned by vulnerabilities. First approximations were made independently at posthunt populations of twenty-animal intervals, and then the results were smoothed slightly. Only a few minor adjustments were necessary because constraints of the interval consistency of the data set quickly indicated if errors were present.

The pronounced shifts in age distribution in relation to population size can be seen in figures 5.3 and 5.4. Average age of the population increases as population size increases. Survival of the first age class for each sex is highly variable. It is low at low population sizes because of the high, relative hunting kill necessary to offset the rapid population increase. As relative kill decreases, survivorship of the first age-class increases at mid-ranges of population sizes, then decreases at higher populations because of mortality of embryos and newborn young. The survivorship of yearlings and young adults shows the same pattern, but the peak is delayed and the older age-classes do not peak at all. Survivorship of all age-classes one and greater tends to converge at K carrying capacity.

The pronounced impact of vulnerability on age structure can be seen in figure 5.3. Females are approximately equal in their vulnerability and show parallel survivorship curves with uniform spacing, while male curves are not so parallel and show pronounced differences in spacing. The population correlates to these shifting age structures will be explored in the following chapters.

CHAPTER 6

Deriving a Recruitment Model

From the data presented in earlier chapters, it is possible to derive models of the population responses to density changes. It is the goal of this chapter to examine the relation of recruitment to population size resulting from past manipulations and to derive therefrom a model to predict recruitment responses to future manipulations.

The complexity of the population dynamics of a large herbivore in the natural world are overwhelming. There is the initial difficulty of determining population size, given the margin of error inherent in most wildlife census techniques. Even when the change is real, it is difficult to determine which one or set of interacting variables produced the change. There is random variation, density dependent effects, seasonality, fluctuations in carrying capacity (including those produced by the herbivore), time lags in herbivore population and vegetation response, human hunting, nonhuman predation, parasites and diseases, and catastrophic events, to mention only the more obvious ones. This situation is not conducive to understanding and mainly accounts for the slow progress in the field. We cannot expect to understand the complexity of interactions without first determining how the population behaves in the absence of complicating variables. Once the simplest case is understood, complexity can be added by determining the effect of each variable independently on the simple case. Finally, interactions of a number of variables can be allowed to occur simultaneously, as they do in the real world. This is simply good experimental design.

It would be absurd to pretend that such control was achieved in the George Reserve deer studies, but in analytical terms, the goal must be to separate the influence of single variables. Only in that manner can we hope to go beyond gross correlation and begin to get at cause and effect. Thus, the approach taken here was to first examine the responses of the deer population in the simplest state possible and then to add complexity to the model, one major variable at a time. In the following chapters the reader should bear in mind that the simplification is by design and does not imply that either the George Reserve or the outside world is so uncomplicated.

The major variables are categorized and defined below. This particular set of variables is arbitrarily selected relative to the analytical problems of the George Reserve deer data. It appears to be the most parsimonious set for extracting the maximum amount of understanding from the specifics of this study. It is not proposed as a general classification since it merges analytical and conceptual components in a particular way, and the particular mix that is useful will vary from study to study.

Independent Controlled Variable. That variable which is used to manipulate the system in order to produce change in the other variables. In this study it is human hunting, the major variable under control which was used to vary population size. It includes both intentional kills and poached kills. Since all known accidental deaths were human related, they are included in this category.

Dependent Variables. Those variables which respond to changes in the control variable.

1. Recruitment. The number of offspring surviving to about six months of age.
2. Prerecruitment mortality. The number of offspring dying between early embryo state and recruitment age.
3. Chronic natural mortality. Arbitrarily defined as that mortality due to factors which debilitate rather than kill directly by injury or trauma. In includes malnutrition, normal burdens of parasites and diseases, old age, etc. It excludes human hunting, nonhuman predation, and virulent diseases.
4. Population size. Total numbers of deer in the George Reserve.
5. Carrying capacity. The basic capacity of the George Reserve to support white-tailed deer. Because of the relative stability of the climate in southern Michigan and slow rate of succession on the reserve, the background carrying capacity has fluctuated little. The major variable influencing carrying capacity has been the size of the deer population. K carrying capacity is defined as the maximum number of deer which can be supported in equilibrium in a steady environment and in the absence of time lags.

6. Time lags. This variable includes time delays in population response to sudden, large changes in population size and the variations in carrying capacity due to time lags in deer response, overshoot of K carrying capacity, vegetation damage, and suppression of vegetation recovery. All of these effects are represented in the George Reserve data set.

Independent Uncontrolled Variables. That proportion of variation due to unidentified factors that are largely independent of the control variable. In this study it is the unidentified variation remaining after the influence of the identified dependent variables that has been taken into account. It includes random effects, errors in techniques, and other unidentified variables.

Simulated Variables. These are variables which were not present on the George Reserve, but which can be simulated from information in the literature.

1. Nonhuman predation. This variable was not present on the George Reserve. Foxes and an occasional badger were the largest predators present. Although they may occasionally succeed in taking a fawn, they function largely as scavengers on deer. Domestic dogs occasionally get through the fence into the reserve, but they are removed as quickly as possible. It is known that deer have been run by dogs, but no mortalities are known to have occurred.
2. Catastrophic events. This variable includes sudden and massive effects of unpredictable occurrence that result in numerous mortalities. It would include exceptionally severe winters, virulent diseases, etc. Frequency will vary with the situation. For example, the frequency of exceptionally severe winters in Canada—the northern extremities of the white-tailed deer range—will be much higher than in southern Michigan, and in that case frequent catastrophes may be the major controlling factor which overshadows all other variables. Unpredictability of occurrence refers not to frequency or probability (e.g., one in ten years) but rather to the fact that the specific years the catastrophe will occur cannot be predicted. Although no catastrophic event has been recorded for the George Reserve white-tailed deer population, the population response to catastrophic events can be simulated.

In the following chapters the analysis will move from the simple to the complex. The simplest case involves the derivation of a deterministic model, i.e., one which lacks variance and treats only the mean population responses in a stable environment. The relationship of prerecruitment and chronic mortality to the simple model is derived next. This is followed by a stochastic model in which variance due to unknown causes is introduced. Time lags are then considered, followed by an examination of nonhuman predation and catastrophic events as they relate to the models derived.

Recruitment as a Sufficient Parameter

The parameters of reproduction and survival are strongly influenced by population density. The question of what, precisely, that relationship is can be approached in two ways. The first involves a factor-by-factor assembly in a complex model where each parameter of reproduction and survival by sex and age is linked to produce a prediction of a final outcome. The second involves integration of the myriad separate parameters into a few, higher order parameters to produce a simple model for prediction. This latter approach is akin to the "sufficient parameter" concept of Levins (1968).

It is the latter approach, using recruitment and posthunt population, that will be followed here; it is simple, straightforward, and the results are consistent with those of a more complex model. In fact, most of the analyses given previously for individual parameters are dependent to some degree on the recruitment data and, therefore, are but different aspects of the same basic data set. As a consequence, even though data may be independently derived from separate sources (as, for example, embryo counts from reproductive tracts or recruitment from reconstructed populations), the final analysis of individual parameters depends upon integration of the two.

A further advantage in using recruitment in modeling is the fact that the deer population is serving as a complex integrator of population responses and environmental influences to yield an outcome—recruitment. Furthermore, the same outcome in recruitment can be achieved by an almost infinite number of combinations of variables for deer and environment. Sex ratio at birth, survivorship by sex and age and resultant age structure, sex ratio of adults, etc., show high variability from year to year. If each of these parameters was treated separately, the variance (which is multiplicable) would be astronomical after only a few variables were entered into the model. Recruitment, which is the functional outcome of the interaction of all of these variables, shows little more variance than any single individual parameter that would go into a complex model. The advantages of the recruitment model in terms of simplicity and lowest variance should be obvious. The approach, therefore, was to derive predictability from the recruitment model and, subsequently, to break down such predictions into more complex parameters.

Recruitment Model

The recruitment model presented here is based upon reconstructed herd numbers for nineteen years. The regression of recruitment on population size is shown in figure 6.1. Since the reconstructed population approached a complete count, it was assumed that the independent variable (x) was measured without error. While this assumption is not strictly true, it is probably close enough to cause little problem with statistical probabilities. Therefore, whenever a reconstructed population number was used as the independent variable, it was treated as a "Berkson case" (Sokal and Rohlf, 1969) and a type I regression model (least squares) was applied.

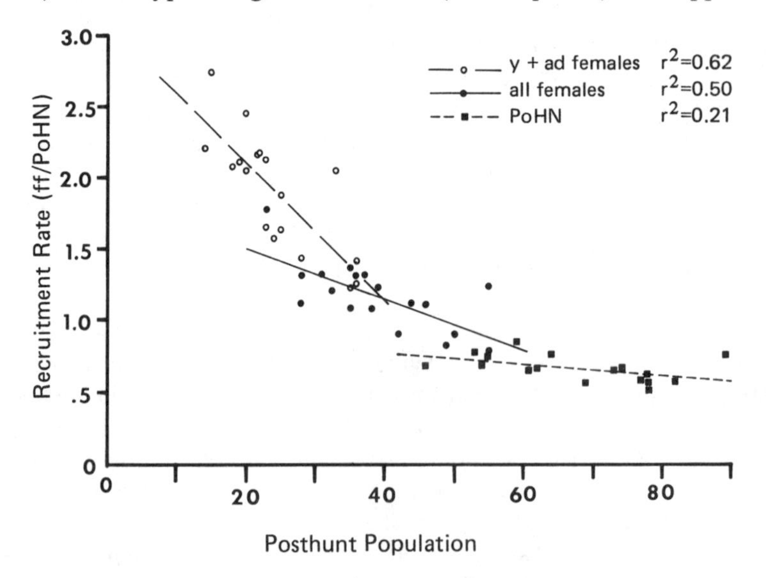

FIG. 6.1. Regression of recruitment rate on the posthunt population of all sex- and age-classes (y = 0.907 − 0.004x), all females (y = 1.874 − 0.018x), and yearling and Y + Ad females only (x = 3.077 − 0.048x)

Regressions were calculated for total population, total females, and Y+Ad females only. Y+Ad females had the highest coefficient of determination ($r^2 = 0.6200$), while total females were somewhat lower ($r^2 = 0.5006$); total population was lowest ($r^2 = 0.2146$). That all slopes were negative showed that recruitment rate decreased as population size increased, a clear demonstration of a density dependent relationship. Similar negative relationships have been found for other big game populations (Gross, 1969), and the analyses by Tanner (1966) suggest they are common for all animals. The fit of the regression is relatively good, considering all of the variables which can influence recruitment.

Ricker (1975) has observed that a frequent criticism of regressions of y/x on x, as used here, is that they are statistically suspect, since such regressions tend to give a negative slope in the absence of a real relationship. However, he further notes that the amount of slope accounted for by the bias is small when compared to the amount caused by a significant variable.

That such a bias does not account for the negative relations in figure 6.1 can be established by using a different analytical approach that does not require the use of x (posthunt population in this case) on both axes. The absolute number of fawns produced can be regressed upon the total females which produced them. The resulting equation is $y = 25.34 + 0.4814x$, and $r^2 = 0.3171$. The lowest observed value of posthunt number of females (23) gives a predicted number of offspring (36.41) which would yield a recruitment rate of 1.58. The highest value of posthunt number of females (55) gives a predicted number of offspring (51.81) which would yield a recruitment rate of 0.9421. Thus, this regression suggests that recruitment rate goes down as posthunt number of females goes up and, therefore, confirms the relationship given in the figure. The relationship can be tested further by the use of a t-test to determine if the slope is different than a given value. If the recruitment rate were constant, a much greater slope would be necessary for even the lowest value, which would relate to the highest posthunt number of females. The observed value was highly statistically different from this closest possible value; and the other values (the $\bar{x}$ for example) were even more different.

If the regression of figure 6.1 for total females is valid, the relationship of number of recruits on total posthunt females should be parabolic, not linear. Plotting of the residual (although there are two outlaying points) demonstrated the existence of the parabolic relationship. It is concluded, therefore, that the regressions obtained in the figure are due to biological causes and are not an artifact of the *y/x* on *x* regression.

Attempts were made to identify important environmental variables that modify recruitment rate. On the basis of deer biology, the following variables were considered to be of greatest importance: total winter snowfall, days of snow cover, summer rainfall (April 1–October 31) and size of acorn crop. Of these, only total snowfall proved to be of importance, and its inclusion in a multiple regression equation increased the R^2 of the total female regression from 0.5006 to 0.5781. Only six years of relative estimates of the acorn crop were available (1966–67 through 1970–71), and they were not correlated with recruitment rate. It is concluded, therefore, that size of the posthunt population is by far the most important variable influencing recruitment rate, and that the environmental variables expected to be important were relatively unimportant. Indeed, most of the unidentified variation in recruitment seemed to be due to random factors or errors in the methods. Presumably, environmental variables would be more important at higher population densities than at those maintained over the last nineteen years.

Examination of Assumption of Linearity

The regressions appear to be approximately linear over the range covered and seem to meet the assumptions, although there is some doubt about the Y+Ad female regression. Analysis of residuals for total population and total number of females gave little reason to question that the assumptions of the regression model had been met, with the possible exception of the time order of the nineteen years. Variance appeared to be somewhat higher in the earlier period of the nineteen years, when relatively great differences in the size of the kill occurred between years (tables 3.1 and 3.2). The possibility of time lags was certainly present. Nevertheless, plotting the residuals against size of kill gave a random scatter. It is concluded, therefore, that no important deviation from the assumptions was present for total females and total population. The question of linearity of the Y+Ad female regression derives from the fact that at lower densities significant numbers of fawn females reproduced, and this might bias the left-hand segment of the regression to the high side. Examination of residuals in this case was suggestive, but somewhat ambiguous because of high variance.

Another question involves the validity of linear extension of the regressions over their full length from *y*- to *x*-axis. It seems likely that the left-hand extension of the *y*-axis is linear for several reasons. First, the extent of extrapolation is relatively short, particularly for the two regressions involving females only. Second, linearity of the recruitment rate is consistent with the derived numbers of young curve (see fig. 6.6) being continuous through the zero intercept, the latter being a necessary constraint. If one assumed that the regression line bent downward on the left-hand end, the outcome would be low recruitments at extremely low densities followed by high recruitment at moderate densities. One could argue for this proposition on the basis of delayed sexual maturity of females. In fact, originally a plateau was expected at the left end of the regression, based upon the assumption of a maximum rate equivalent to the intrinsic rate of increase (r_{max}). This assumption came from the expectation that there was a physiological maximum limit on the number of young that could be conceived and born. Intuitively, it seemed that this capacity could not increase without limit. Such a plateau does not exist, however, because pregnancy rates of Y+Ad females continue to increase and fawns begin to breed in their first year of life. Data from table 4.4 and figure 4.8 show that the embryo rates of female fawns and Y+Ad females continue to increase over the entire span on the left-hand end of the regression line. Hence, it is concluded that for all practical purposes, the linear extension of the regression to the *y*-axis is correct.

Extension of the regression on the right-hand side of the *x*-axis is more problematic. At present, the nineteen years of reconstructed population data cover breeding population densities to just beyond the midpoint of values, assuming that the regression line is linear. Therefore, estimates of maximum sustainable yield (MSY) can be derived at this time, since MSY will be at or close to zero recruitment divided by two if the recruitment rate on population size regression is linear. However, there is no direct evidence from the reconstructed population data to test the question of linearity in the right-hand direction. However, if deviation from linearity occurred, one would expect that it would be in a downward direction, because as density of deer increased, competition for resources would intensify. Since the range covered by the present data points includes some influence of competition, deviation from linearity would require an accelerating increase in competition.

Densities at the high extreme were achieved in the early years of the George Reserve deer population, and data derived from the drive counts during this period can be examined with respect to the question of linearity. Drive counts were corrected, as explained in chapters 2 and 3.

The procedure of determining the recruitment rate from early drive count data was as follows: (1) the corrected drive count gives the prehunt population; (2) the total removal is subtracted from the prehunt population to obtain the posthunt population; (3) the posthunt population is subtracted from the subsequent prehunt population to arrive at recruitment; and (4) the recruitment rate (as with the rate of the reconstructed population data) is obtained by dividing the recruitment number by the posthunt population of the preceding year. Several comments on this procedure are necessary for its evaluation. The corrected drive count is the source of error in the data. An earlier discussion of the drive census method pointed out the problems involved. The removal, however, is a known value. Since removals usually were directed towards producing a predetermined population size, they tended to introduce a stabilizing factor. For example, assume an erroneously high drive count was obtained; then, an exceptionally high kill was taken, which would depress the actual population considerably. The probability is that the lowered actual population in the following season would yield a low drive count and a low removal. The effect of the interaction of the known value with the unknown value is to reduce absolute error in the unknown. Extreme errors in one year tend to be balanced by opposite errors in the following years. The outcome has high variance, but over several years it gives a reasonably accurate estimation of the true mean.

Given that there is a high variance between years but that the mean over several years is a reasonable estimate of the true mean, a method of smoothing is required. A five-year running mean was used, with the data point being the midpoint year in the five-year series, except for the beginning and end of the forty-three-year series.

Although the resulting plot (fig. 6.2) shows several idiosyncracies, the overall impression is one of essentially linearity. The relationship of recruitment rate on posthunt population size was negative, which agrees with the reanalysis of the early George Reserve data of O'Roke and Hammerstrom (1948) by Gross (1969). Therefore, it seems reasonable to assume that the linear extension of the regression in figure 6.1 is reasonable.

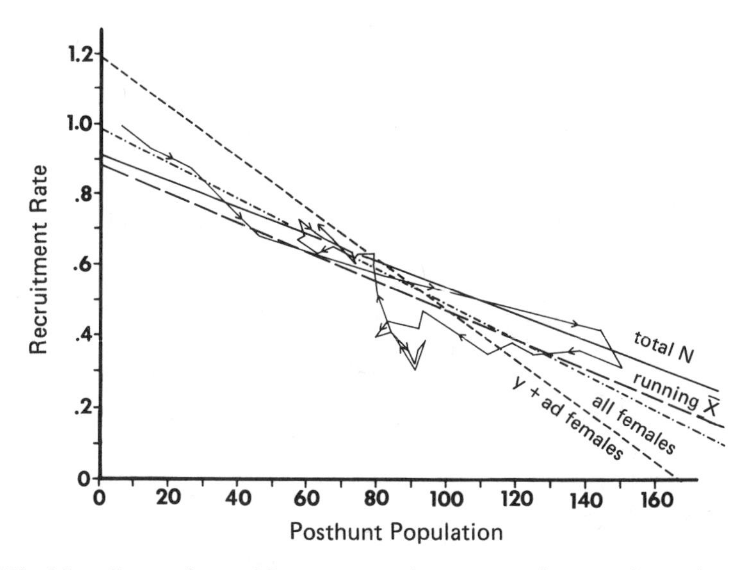

FIG. 6.2. Comparison of five-year running mean estimates of recruitment rate (arrows indicate time sequence) on posthunt population size with regressions of the running mean, total posthunt population, all females, and yearling and older females only

Relationship of Recruitment to Sexes of the Posthunt Population

Another unexpected outcome of the regressions of recruitment rate on population size is that the regression fit was better for the female segment of the population alone than for the combined males and females (fig. 6.1). On first thought, one might conclude that the females should be closer since they are the producers of young, and the exclusion of males would eliminate variation due to variable sex ratios in the population over the years. But recall that the regression is *negative*. As the number of females goes up, the fawns recruited per female goes down, and beyond a certain point (MSY) the absolute number of fawns recruited goes down. Thus, it is a strongly density dependent response which contains elements of both declining conception rates and declining survivorship of young. As one moves from left to right along the regression, the importance shifts from the former to a combination of both factors. Competition for resources becomes the critical issue, and this conclusion is reinforced by the concommitant variations in growth and other weight responses.

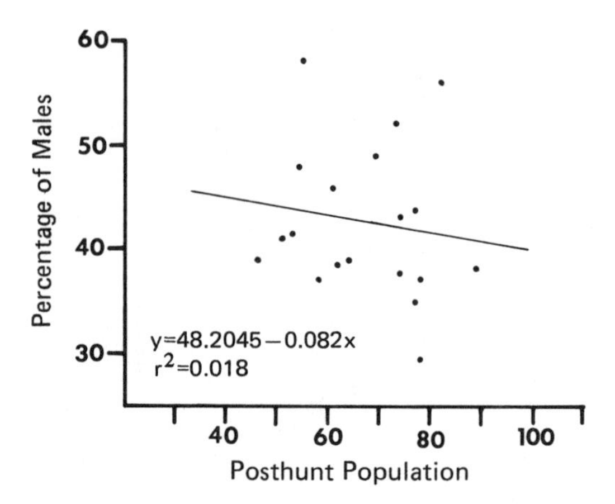

FIG. 6.3. Regression of the male percentage of the posthunt population on the size of the posthunt population

The proportion of the posthunt population which is male and female has varied considerably and unpredictably over the nineteen years of reconstructed population. Figure 6.3 shows a plot of the percentage of the posthunt population that was male against the size of the posthunt population. The slope is essentially zero (b = 0.0820). Furthermore, the scatter is virtually random, as indicated by the low r^2 (0.0183).

Furthermore, changes in sex ratio do not account for variation in the recruitment rate regression. In figure 6.4, the deviation of observed recruitment from expected (i.e., that indicated by $\hat{y}$ of the regression at a given value of x) is plotted against the percentage of the posthunt population that was female. If sex ratio were important, increasing the percentage of females should result in recruitments greater than expected, and the slope should be clearly positive. In fact, the fit is extremely poor (r^2 = 0.0408) and the slope is far from being significantly different from zero (p = 0.40). Moreover, if recruitment rate is regressed on the sum of the males only, a very low r^2 (0.1280) is obtained. It appears, therefore, that there is little correlation between the recruitment and the number of males in the population, and what "fit" there is with total population is due primarily to the females, rather than the males. Sex ratio of the posthunt population has little influence on recruitment of young.

At first it would seem that this result could be explained by the fact that at lower population sizes (as was the case for most of the period considered in these regressions) virtually all of the offspring

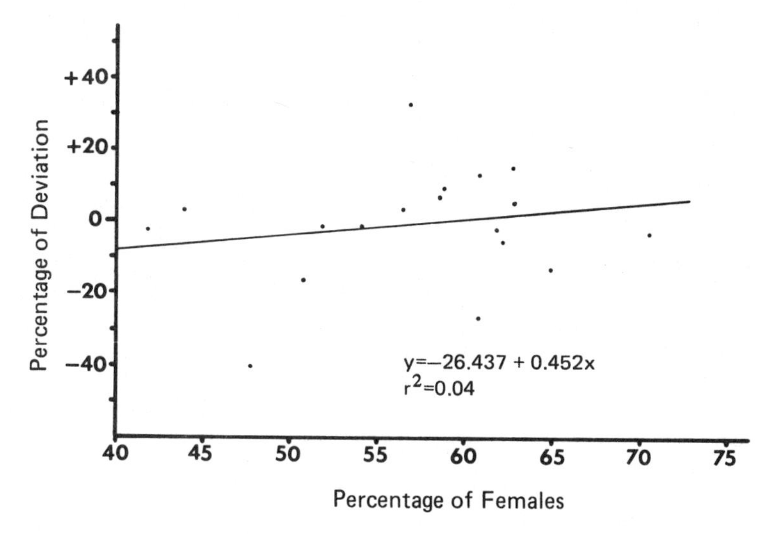

FIG. 6.4. Relationship of the deviation of observed recruitment from expected to the female percentage of the posthunt population

produced survive. But recall that embryo rates also varied with density, and since virtually all of the embryos survived to recruitment, the correlations of embryo rate would have been with females only, not total population. Thus, the basic question remains.

Given that competition is the important element, should not the total number of animals be more closely related to the number of young recruited? If food is in short supply, it would seem to make little difference if it were depleted because it was consumed by males or females. Why does the number of young recruited track more closely the number of females in the population than the total number of animals?

Two alternative explanations are available. The first is that some factor other than competition underlies the negative relationship of recruitment and population density. This explanation seems untenable from all lines of evidence. The overwhelming outcome of results on population responses, growth and weight of deer, quality and quantity of vegetation, etc., points to intraspecific competition for food as the fundamental force behind the observed density dependent responses. Furthermore, the only apparent alternative to competition would be social regulation, and on the basis of both observed population responses and social behavior this alternative is untenable.

The second explanation would be that competition between the sexes is not equivalent to competition within a given sex, and, therefore, the effects of competition are not additive irrespective of sex. Niche segregation by sex based on sexual dimorphism has been suggested for birds (e.g., Selander, 1966; Storer, 1966; Earhart and Johnson, 1970; and others). The concept has been little applied to mammals, even though sexual dimorphism and spatial segregation by sex are common and the same arguments could apply. Differential niches by sex in the George Reserve could account for the apparent contradiction raised by the regressions of figure 6.1. Further discussion of the possibility of niche separation by sex in George Reserve deer will be given in chapter 14.

Selecting the Best Regression for the Recruitment Model

It is desirable to have the best predictor for modeling purposes. On the basis of fit, the regression with the highest r^2—the Y+Ad females—would appear to be the best choice. In addition to a close fit, however, the accuracy of the slope (b) is of fundamental importance. An inaccurately estimated slope will result in extreme errors of estimation at the ends of the regression line. All three regressions of figure 6.1 will give similar predictions in the middle range of the regressions.

The accuracy of the estimation of slope can be tested by comparing the slopes of the nineteen years of reconstructed data with that of the drive count data with the greater range of population sizes represented (fig. 6.2). Several criteria can be used to judge the regression which best estimates the actual slope. First, it must transect the data points based upon the nineteen years of reconstructed population. Of course, all three regressions of figure 6.2 do, since those are the points from which these regressions are calculated. The only point of bringing up this criterion is to reject another possible regression, that based upon the points of the five-year running mean for drive counts. It can be seen that this regression clearly fails to transect the nineteen-year data points. Although the regression has a good fit ($r^2 = 0.6303$), it is partially due to greater sample size ($n = 43$), broader range of x values, and, particularly, the fact that much variations has already been masked by using the five-year running mean. Therefore, this regression has been rejected for further modeling.

Close examination of the plot of the five-year running mean in figure 6.2 and the population size in figure 3.1 reveals a long early period of population growth during which the recruitment rate declined, even though it remained higher than expected at the peak populations. This was followed by another long period of gradual reduction of the population size, during which time the recruitment rate was below that expected. Finally, control of the population size to moderate levels resulted in the cluster of points obtained during the nineteen years between 1952–53 and 1970–71. An interpretation of this record follows.

There was a long period of rapid population growth because of low population density and little competition for food (1928–29 through 1935–36). The unexpectedly high recruitment rates at the peak population were due to a time-lag effect. The absence of deer from the area of the George Reserve resulted in an accumulation of food resources that was used by the deer population to sustain population growth beyond the expected equilibrium level. This population "overshoot," in turn, depressed the vegetation growth of preferred deer food species. Thus, a second time-lag effect of a negative sign was created, resulting in depressed recruitment rates during the period of herd reduction (1936–37 through 1950–51). Finally, the gradually lowered population levels allowed vegetation recovery, and recruitment rates returned to average levels (i.e., those achieved in the absence of time lags—1952–53 to 1970–71). This explanation seems reasonable, logical, and consistent with what is known about ungulate population-vegetation interactions. Furthermore, it is consistent with what has been written about the history of the deer herd and vegetation of the reserve. Finally, it is consistent with the other available data of this study.

Assuming that this depiction approximates the true course of events, a second criterion becomes available. The best regression is one that lies intermediately between the extreme points of the five-year running mean at the high population density. It should not pass through the points produced during the population overshoot time lag, nor should it pass through the points of the deteriorated range time lag.

The third and last criterion is that the regression should reasonably approximate the early initial growth of the deer herd. As discussed in chapter 3, the early growth of the population can be approximated by a projection based upon independent data derived from the examination of reproductive tracts. Also, this projection implies the occurrence of an "overshoot" time lag. Thus, the best regression, for a time lag free model, also would lie below the five-year running mean points of the peak population period.

The three regressions of figure 6.2 can be evaluated by these three criteria. The regression with the best fit, the Y+Ad female regression, would seem to overestimate population growth at the left end and underestimate it at the right end. The steepness of its slope appears to be due to a higher cluster of points at the left-hand end of the nineteen-year reconstructed population data (fig. 6.1), relating to questions about the assumption of linearity of the regression over the range of values represented. It appears that the slope of the right-hand cluster of points is influenced primarily by reproduction of Y+Ad females, while the upsurge at the left-hand cluster is due to the onset of breeding of fawn females. The breaking point for the significant amount of breeding by fawn females (table 4.4) would seem to lie suspiciously close to the possible segregation of points in the Y+Ad female regression in figure 6.1. Such findings resulted in its elimination from further considerations.

The total population regression underestimates the early population growth and approaches the points of the overshoot period. Thus, the predictions on both ends would seem to be in error. The errors on the left end would probably be minor, but those on the right end would be excessive. The zero recruitment value of 245 deer exceeds the peak population of the initial growth phase. Because this regression yielded unsuitable data, it too was eliminated from further consideration.

The total female regression seems to be the best choice. It does not raise the same questions about linearity as the Y+Ad female regression since fawn females are included on the x-axis. It most closely approximates the early growth period and splits the difference on the right-hand end, as evidenced by its closeness to the five-year running mean regression. Also, it has a reasonably good fit. Therefore, it was the regression selected for further population modeling.

The regression of all females on recruitment rate, extended to the x- and y-axis and with 95 percent confidence limits, is given in figure 6.5. Since males would eventually have to be added to complete the breeding population, they are added to the x-axis at this time as 47.74 percent of the population. They are incorporated now so that they may be used in further manipulations and to avoid having to add them to separate calculations. The end result is the same, but their inclusion simplifies the procedure.

The zero recruitment point (where the regression intercepts the x-axis and recruitment rate = 0) is 198.11, and the MSY level (that breeding population which yields the highest possible recruit-

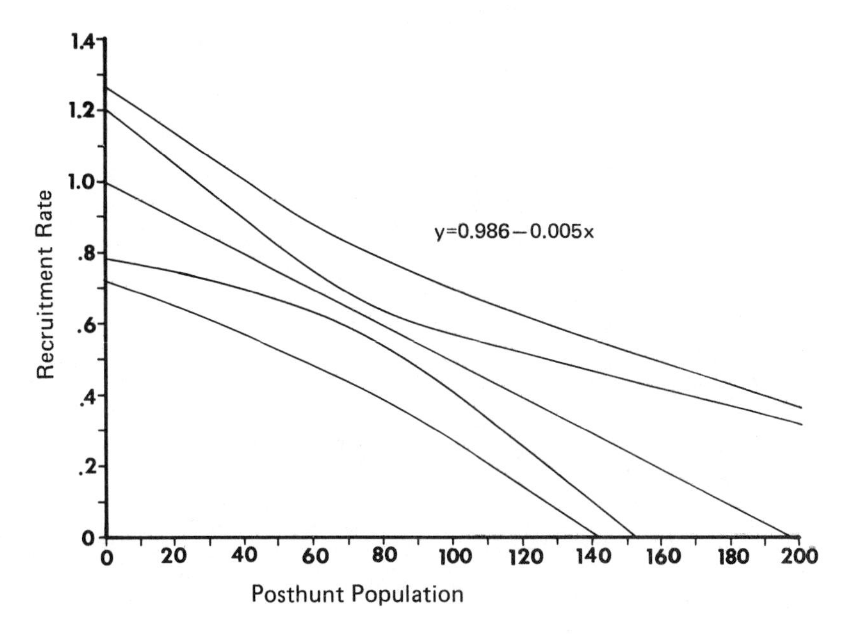

FIG. 6.5. Regression of recruitment rate on posthunt population size. Inner band shows 95 percent confidence limits of the mean of y at a given value of x; outer band shows 95 percent confidence of the individual value, y.

ment of numbers of offspring) as given by zero recruitment divided by two is 99.05. The maximum rate of recruitment (i.e., the y-axis intercept) is 0.9868. The recruitment in numbers is obtained by multiplying the recruitment rate (y) by the breeding population (x) of this regression. The recruitment of numbers in relationship to the size of the breeding population is shown in figure 6.6. Since the regression of recruitment rate on posthunt population is linear, the numbers recruited form a symmetrical parabola in which zero recruitment of numbers is identical to zero recruitment rate (198.11). The MSY occurs at the peak of the parabola, where a posthunt population of 99.05 gives a recruitment number of 48.69 young. This is the mean maximum number of young that the population can recruit and is the best possible trade-off between breeding population size and recruitment number, which declines as population size increases or decreases. The declining right-hand segment of the parabola represents failure of survivorship of young born because of lack of resources; the left-hand segment represents high survivorship of offspring, but a declining number of females in the

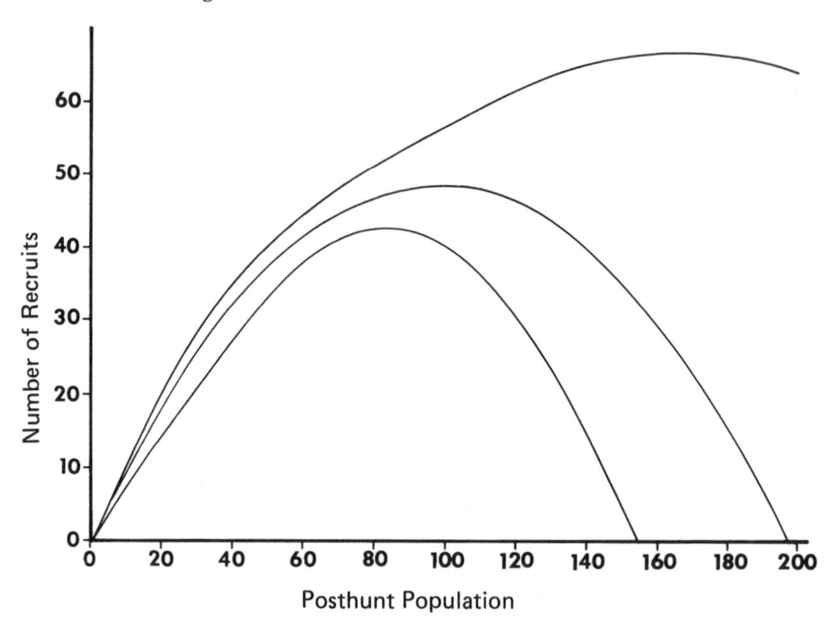

FIG. 6.6. Parabola of numbers of fawns recruited with 95 percent confidence limits. Maximum recruitment of numbers is 48.69.

breeding population. Although the breeding rate increases, it does so at too low a rate to offset the decline in the female breeding population. This segment of the curve, of course, must pass through the zero intercept (no breeding population, no offspring).

From the left-hand segment of the parabola, it should be intuitively obvious why the left-hand extension of the recruitment rate regressions of figure 6.1 were considered to be linear. Given that the middle range of the recruitment numbers parabola is strongly established by data points and that the parabola must pass through the zero intercept, the number of possible shapes of this line are highly constrained. A depressed regression line of recruitment rate, due to a leveling of reproductive capacity, would not alter the middle range of the parabola but would yield a straighter line from the zero point to the mid-range. A completely straight line connection would imply that all females reach an upper limit of reproductive rate, an assumption that seems unlikely in view of the continual increase in embryo rates and greater proportions of twins and triplets as the population size becomes smaller. Therefore, the assumptions of linearity in extending the regressions in figure 6.1 to the y-axis seem correct.

The left-hand segment of the parabola of number of recruits is a stable area for reasons other than correctness of extrapolation. The amount of variance is quite low, as indicated by the narrow confidence limits in figure 6.6, and reflects a real biological phenomenon. Virtually all of the population response is accounted for by reproduction, since nearly all offspring survive. These low population densities result in relatively little resource competition, and fluctuations in the environment have relatively little influence. Finally, time-lag effects, which are most pronounced at the K end of the scale because of overshoot and depression of the resource base, are virtually absent. The result is that the left-hand segment is a model with good accuracy and high predictability.

The right-hand segment of the parabola is a morass of unpredictability, as the confidence limits become extremely wide. Part of the problem, such as the question of linearity of projection of the regression of recruitment rate to the x-axis of breeding population, is a characteristic of the model, but most of it is a reflection of the real state of the natural system. This is because most of the response is not dependent on reproduction, but on survivorship of young, and the response becomes increasingly more weighted to survivorship as one moves toward higher populations. At this end of the regression, recruitment is nearly independent of reproductive effort (see fig. 4.5). It is primarily dependent upon competition for resources. Hence, natural fluctuations in the resource base due to rainfall and other climatic factors, or fluctuations in fruit and mast crops and other valuable resources would be expected to have a profound effect upon annual recruitment. Time-lag problems become pronounced, and even the length of a time lag becomes an important variable. For example, once overshoot and depression of the resource base occurs, such a depression can be maintained almost idefinitely if the population is not greatly reduced to allow for recovery. Furthermore, catastrophic effects are more extreme at high population densities. For example, a hard winter often has little effect upon a low population, while severe die-offs occur when the population is at the limit of resources.

The strategy followed in this study has been to eliminate as many variables as possible to obtain data on the basic response of recruitment to density. Thus, the relative stability of the habitat on the George Reserve is a plus. Furthermore, time-lag effects have been avoided by taking an intermediate regression for the model, a regression that splits the difference between the two, directionally opposite time-lag effects of figure 6.2. Another justification for this approach is that time-lag effects can be dampened or eliminated by management directed towards introducing stability.

CHAPTER 7

Converting the Recruitment Model to a Population Model

The recruitment curve in figure 6.5 considers only recruitment with a given posthunt population. It does not indicate the mortality of older animals. Therefore, the zero recruitment point (i.e., the x-intercept) will not give an accurate estimate of the true population equilibrium at K carrying capacity. That equilibrium will occur at some point below zero recruitment, where the chronic mortality of older animals equals the recruitment of young. Thus, in the absence of time-lag effects, a posthunt population giving zero recruitment would never be achieved.

At low posthunt population sizes, chronic mortality can be ignored, since hunting kill essentially eliminates chronic natural mortality and an equilibrium is achieved through management. Under heavy shooting programs, all of the mortality of older animals can be accounted for by shooting. For the period covered by this study (where cause could be determined), all the aminals found dead had suffered gunshot wounds. Furthermore, the oldest individual found dead was considerably under the natural, maximum longevity of the George Reserve population. The oldest animal found dead was nine years old, and most were substantially younger. The age of the deer and the totals found dead, from older to younger, are: seven years, one; six years, one; five years, two; four years, five; three years, eight; two years, seven; one year, twenty; and young of the year, thirteen.

The maximum longevity of George Reserve deer is approximately thirteen years. The oldest animals shot were twelve and one-half years old (tables 3.1 and 3.2), and if they had not been shot, presumably they would have lived into the following year. Maximum longevity is set by the wear on the cheek-teeth. On the

George Reserve, about twelve to thirteen years is the time at which the crown is worn away on the lower first molars, with only the root stubs remaining. Abscessed jaws and ineffective mastication limit further life expectancy.

Thirteen years is the ecological, rather than record, longevity. Older deer are known from the wild (Ryel et al., 1961; Ozoga, 1969), and an occasional animal on the George Reserve will live beyond thirteen years. But some will also have worn out teeth at an earlier age, and thirteen years seems to be the typical longevity. This is greater than the longevity recorded from the Seneca Army Depot in New York by Hesselton et al., (1965) and for New York state by Maguire and Severinghaus (1954). Ozoga reported that of almost two thousand deer marked in the Upper Peninsula of Michigan, no males and only three females exceeded thirteen years of age. The oldest female was fourteen years, nine months. Thus, the estimate of the ecological longevity of the reserve deer population seems reasonable.

As shooting declines (as in the current phase of management), the chronic natural death of breeding animals will increase because of density dependent effects. Thus, the fewer the animals removed by shooting, the greater the number that will die of chronic causes or old age, if other hazards are survived. Therefore, at higher population levels (i.e., those above MSY), the recruitment curves are not a satisfactory population model. They cannot predict either the combination of limited shooting and natural mortality at a given posthunt population size or the K carrying capacity, where chronic mortality alone comes into equilibrium with recruitment.

To produce a satisfactory population model, the right-hand segment of the recruitment curves must be changed from "recruitment" to "population change" by shortening the distance on the x-axis, where the curve crosses the axis. Under no hunting, the point where the curve crosses the x-axis should be equilibrium K carrying capacity (where the mortality of older animals by totally chronic causes is balanced by recruitment). The deviation of the population change curve from the recruitment curve should occur at that point where hunting is light enough that some animals escape and live long enough to die of chronic causes.

Estimating K Carrying Capacity From Survivorship

At the present time, the value of the points where chronic mortality begins and K carrying capacity is achieved cannot be determined directly. The next phase of the study, involving maintenance of high population densities, is specifically designed to obtain these answers. On the basis of present information it can be determined that the two points are between 90 (the highest breeding population observed) and 198 (the recruitment zero point). On the basis of age structure, the values can be estimated somewhat more precisely. These estimates are based primarily on life expectancies in relationship to maximum longevity.

Figure 7.1 shows the relationship of the total breeding population to the number of female recruits, as predicted by figure 6.6 and corrected for sex. Only females are used, since they are the population producing the recruits and whose life expectancies will be the major influence on the number of breeding females in the population at K carrying capacity. The life expectancy curve of the figure shows the mean life expectance of female recruits (note that this expectancy is from recruitment age, not birth) assuming population equilibrium (table 7.1). In other words, the average female

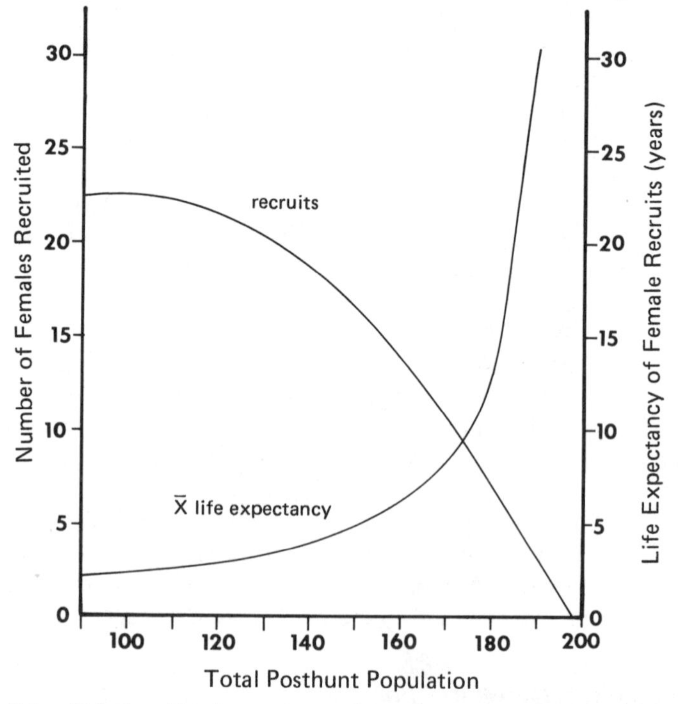

FIG. 7.1. Relationship between total posthunt population, number of female fawns recruited, and required average life expectancy of female recruits to sustain equilibrium of the total posthunt population

TABLE 7.1. Mean Equilibrium Values for the George Reserve Deer Population

Posthunt Population	*Posthunt* ♂♂	*Recruits* ♂♂	*$\bar{x}$ Life Expectance*[1]	*Posthunt* ♀♀	*Recruits* ♀♀	*$\bar{x}$ Life Expectance*[1]
100	47.73	26.00	1.84	52.27	22.69	2.30
110	52.50	25.66	2.05	57.50	22.39	2.57
120	57.28	24.79	2.31	62.72	21.63	2.90
130	62.05	23.38	2.65	67.95	20.40	3.33
140	66.82	21.44	3.12	73.18	18.71	3.91
150	71.60	18.97	3.77	78.41	16.55	4.74
160	76.37	15.96	4.79	83.63	13.93	6.00
170	81.14	12.42	6.53	88.86	10.84	8.20
180	85.91	8.34	10.30	94.09	7.28	12.92
190	90.69	3.73	24.31	99.31	3.26	30.46

[1]Life expectance is from recruitment age, rather than from birth.

must live this long following recruitment in order for equilibrium to be sustained. Thus, at a total breeding population of 170, there are 88.86 breeding females which produce a total of 23.26 recruits, 10.84 of which are female. For 10.84 recruits to maintain a breeding population of 88.86 females, the average recruit must live 8.20 years (i.e., 88.86 divided by 10.84 equals 8.20).

From the figure and table it can be seen that an upper limit on population growth will be set by longevity. For instance, a total population of 190 would require the average female recruit to live 30.46 years, a clear impossibility. If all females lived to maximum longevity (13 years), the population would stabilize at a maximum of approximately 180 deer. Again, it is highly unlikely that all females would live so long. Some would be bound to die at an earlier age due to natural injury, disease, etc., exacerbated by high population density and attendant resource competition.

An estimate of the minimum value of K carrying capacity can be derived by assuming that mortality occurs at a constant rate on all age classes of females following recruitment. The rate need not be strictly constant, but only that females older than the age of mean life expectancy not have a higher survivorship than those younger than the mean life expectancy. This situation is shown in figure 7.2. Line *A* is a straight line that begins with a point of origin of recruitment on the *y*-axis and passes through the mean life expectancy for that recruitment, intercepting the *x*-axis at thirteen years. The combination of 12.75 recruits and a 6.38-year mean life expectancy occurs at a population of approximately 163. Thus, the

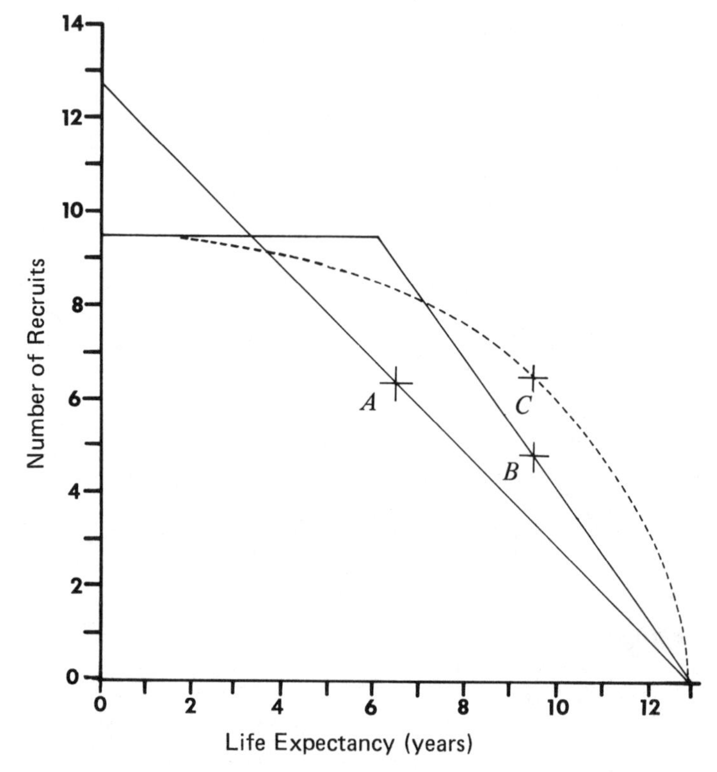

FIG. 7.2. Relationship of maximum number of female recruits and further female life expectancy. The crosses indicate mean life expectancy.

minimum value of K carrying capacity can be set at 163, and the actual value now has been constrained to between an absolute minimum of 163 and an absolute maximum of 180.

The highest possible equilibrium point approximates the intercept of recruits and life expectancy, or approximately 174 total population. At this point, 9.5 recruits would live an average of 9.5 years. If the population reached a higher level (due to a particularly good forage year, for example), it would tend to return to the intercept point in following years, since life expectancy is not sufficient to mainatain the higher population given that the average recruitment continues to decline with population increases. Therefore, 174 is close to the average upper limit of population growth, and K carrying capacity cannot greatly exceed this value on the average.

The question of whether the population achieves this level is more difficult and hinges upon the likelihood of females being able to sustain average life expectancies of 9.5 years, given that the mean maximum is only 13 years. If, for example, the mean life expectancy under no hunting was 7.5 years, the population would stabilize at this point on the life expectancy curve, or at a population of 167. Therefore, while the subsistence population could not greatly exceed 174, it could have a lesser value where average recruitment and average life expectancy (due to chronic causes) balanced each other.

Curve *B* of figure 7.2 is one curve which is consistent with a population of 174 and a mean survivorship of females of 9.5 years. The curve seems unrealistic, since animals would have to have perfect survival until 6 years and then die at a constant rate. From what is known about the subsistence populations of large ungulates, it seems highly likely that recruits survive well during their prime young adult life and die at an accelerating rate as they reach old age (e.g., Deevey, 1947; Caughley, 1966). A whole family of curves of this general shape could be generated as long as the y-intercept is 9.5, the x-intercept 13, and the mean life expectancy 9.5 years. Curve *C* is one of this family and approximates the expected form.

Curves *A* and *C* of figure 7.2 are plotted as life-table survivorship curves in figure 7.3. Although curve *C* represents a strong case of survivorship to maximum longevity, it seems to be a realistic depiction. Since adult animals survive at the expense of recruitment, one would expect the population, as it grows toward K carrying capacity, to shift from *A* to the direction of *C*, with its lower recruitment and greater survivorship of adult animals. In an ecological context, this means that adults are the most effective competitors for limited resources and that juvenile animals cannot meet this competition successfully. Also, because of competition, few females have sufficient resources beyond those necessary for their own maintenance to successfully produce young. In an evolutionary context, it means that each individual female is attempting to maximize her own survivorship—because living to engage in further reproductive efforts is a better strategy than sacrificing herself for the production of a given offspring. The universal trait in large ungulate populations near K carrying capacity is to have old-age structures and low recruitments. The result is that, in the absence of time lags or catastrophic events, population equilibrium occurs when decreasing recruitment comes to balance with the chronic mortality of adults. In seasonal environments catastrophic events are common, while in environments where seasonality is less pro-

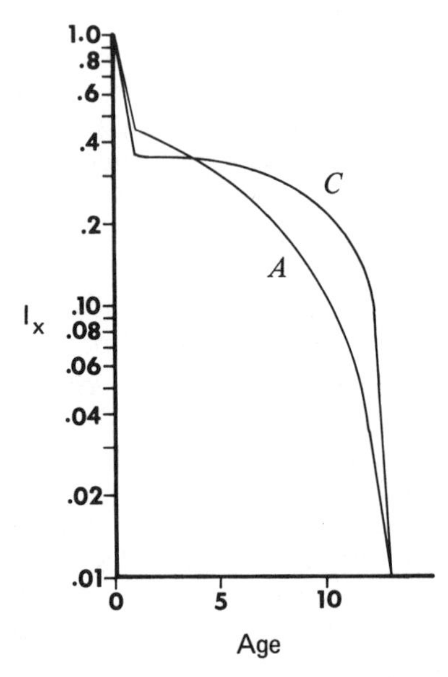

FIG. 7.3. Survivorship curves for females from birth, as derived from embryo rates and survivorship

nounced, subsistence populations at K carrying capacity equilibrium are approximated.

Although these analyses have been based upon survivorship of females, the results would have been similar if based upon males only (table 7.1) or on the average of both sexes. Comparisons of recruitment rates of the sexes and the equilibrium values of expectancy are shown in figure 7.4. It can be seen that the curves are similar, and maximum population sizes derived from one sex or the other would show minor differences.

All of the calculations were based upon a fixed, recruit sex ratio of 1.15 males to 1 female (46.59 percent female). However, figure 4.9 shows that the primary sex ratio becomes unbalanced in favor of males as population density increases. Differential mortality did not occur over the range in population densities covered by the data for nineteen years, but at the low densities present, virtually all fawns survived. (At higher population densities fawns begin to die as recruitment declines.) Differential mortality, with males dying more frequently, is common in ungulates (Taber and Dasmann, 1954; Robinette et al., 1957; Flook, 1970) and would tend to direct the primary sex ratio, which is shifting towards males at

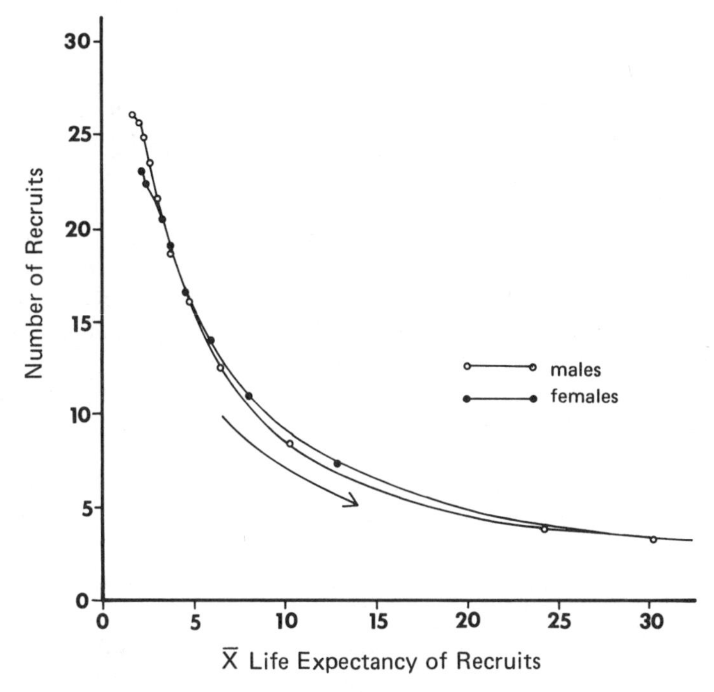

FIG. 7.4. Relationship of recruitment to mean life expectancy by sex at equilibrium population (as derived in table 8.1). Arrow indicates directional tendency in the absence of a hunting kill.

higher population densities, back to the observed ratio at recruitment age. If, however, mortality of fawns by sex were balanced at high population densities, a relatively lower proportion of females would be recruited, thus shifting the estimate of K carrying capacity downward. The calculations of K carrying capacity were repeated with the changing sex ratio to evaluate the magnitude of this factor on the outcome. The effect was trivial, shifting the maximum estimate of K carrying capacity from 174 to 172. Therefore, the original estimate of 174 was retained.

Estimating Population Size at Which Chronic Mortality Begins

Having derived an estimate of K carrying capacity, it is now necessary to determine the breeding population at which chronic natural mortality first becomes important. At the lower population levels of the nineteen years covered by this study, artificially induced mortality by shooting supplanted chronic natural mortality. No deer's death could be attributed to a cause other than gunshot wound (or, in one case, an arrow wound), and only one individual (age nine) even approached the maximum longevity of thirteen years. It would be absurd to maintain that animals never die of natural causes under heavy shooting regimens, since the possibility of an occasional animal escaping hunting throughout its life span is clearly present. But, as a population phenomenon, such occurrences were trivial under the management of these nineteen years. As shooting removals decrease and the breeding population increases, at some point chronic natural mortality will begin to occur regularly. An estimate of that point is important, since it is there that the population model curve will deviate from the recruitment curve (fig. 6.5) and pass through K carrying capacity.

Constraints can be set on the posthunt population level at which chronic mortality will begin. Levels up to approximately 90 animals have been observed, so the point at which chronic mortality begins must be greater that 90. Similarly, an upper limit can be set at a total population of 163 with an average kill of 27.4 animals per year. At that point, if all age classes were subjected to an equal probability of being shot, all animals *could* be killed by the age of 12.5 years. At higher total population levels with lowered kills, some individuals would have to live longer than 13 years, which is considered beyond the longevity of the animals. Thus, chronic mortality would have to occur. Again, it should be emphasized that this would be true only if all age-classes were equally vulnerable to shooting. This was shown not to be the case (fig. 5.2), since some animals, particularly yearlings, were far more vulnerable than average. If the kill were completed with a take of predominately young animals, some older animals would be escaping to die of chronic causes. Therefore, the realistic posthunt population size where chronic mortality would begin is substantially below a 174 posthunt population and considerably above the 90-animal minimum. These numbers should be considered the extremes possible, rather than the likely estimate.

Thus far we have been concerned with mean values as estimators, but the estimate of the minimum population density at which chronic mortality occurs involves a large element of chance. As the number of animals removed from the population declines, the probability of a few individuals escaping and living to the maximum longevity increases. Furthermore, as the population density increases, the probability of other chronic mortality before old age also increases. Therefore, the point at which shooting first fails to include all mortality will be set by the extremes, rather than the means. Obviously, nonshooting mortality can occur by chance at even extremely low densities. The concern, here, is to determine a posthunt population size at which the probability of nonshooting mortality becomes significant, and not just a chance occurrence.

As shooting is reduced, population size increases and life expectancy goes up. Chronic mortality begins at the level at which the oldest animals in the population reach the longevity of the population (i.e., thirteen years) without being shot. Thus, this point can be estimated by regressing the oldest aged animal in the population against posthunt population size. The regression line gives the mean oldest age, and the confidence limits give extremes at various levels of probability. It is the latter that concerns us.

While we have been interested in the means, we have emphasized the female segment of the population, since mean survivorship of females is greater than males. However, such is not necessarily the case with the extremes. The regression of the oldest female in the population against posthunt female population, as expected, was approximately parallel to the same regression for males but was one age class greater (female regression, $y = 3.3887 + 0.1275x$; male regression, $y = 2.3297 + 0.1413x$) at a given population size. Based upon the means, therefore, females would begin to die of chronic causes at a lower posthunt population size than males. However, the fit of the male regression equation was substantially poorer than that of females ($r^2 = .1039$; female $r^2 = .4460$) and greatly influences the confidence limits. For females, the upper 95 percent confidence limit of the $\hat{y}_x$ crosses the thirteen-year-old age-class at a female posthunt population of 55, which is equivalent to a total, male and female posthunt population of approximately 105. For males, crossing occurred at a posthunt male population of approximately 37, which is equivalent to a total posthunt population of approximately 71 head. If one trusted these results, the conclusion would be that males die of chronic causes at a lower population size than females. However, this is an artifact of the high variance of the male regression. The estimate of population level at which chronic mortality for males begins seems far too low and is inconsistent with the observed data on the George Reserve, where posthunt populations exceeded 71 by a considerable margin without an observed case of chronic mortality. Nevertheless, it emphasizes the need to consider both sexes in terms of estimating the likelihood of reaching longevity.

Consequently, the oldest individual in the population, regardless of sex, was regressed against total posthunt population. The resulting equation ($y = 2.2986 + 0.966x$; $r^2 = 0.3377$), with its associated confidence limits, was used to estimate old-age mortality according to the following rationale. Each point of the regression represents x and y values for a given year of the nineteen years of the study. Thus, each point represents one year as well as a given value for x and y. The assumptions of the regression equation are that each y-array on a given value of x is normally distributed, and the variances of the arrays are equal. These assumptions seem to be reasonably well met by the data. One can then determine the posthunt population size which falls on the 95 percent confidence limits of $\hat{y}_x$ at age thirteen. In this case it is 79. Now assume that the posthunt population were held by shooting to 79 for one hundred years. The probability is that in five of those years, one animal will reach the age of thirteen (or escape over twelve harvests) and die of old age. Thus, by these probabilities, 5 deer die of old age in one hundred years. However, in the one hundred years there were 7,900 posthunt deer which did not die of old age. Therefore, the percentage of old-age mortality can be calculated as 0.06 of 1 percent.

By changing the probability level (based upon t distribution values with n-2 degrees of freedom, in this case 17) to 90, 80, 70, 60, etc., and repeating the above procedure, one can generate a series of probabilities of natural death dependent upon posthunt population size. Regression of the percentage of natural death on posthunt population size gives a linear relationship according to the equation $\log_{10} y = 13.4494 + 6.4605 \log_{10} x$ ($r^2 = 0.9918$). With this equation, the chronic natural mortality rate can be predicted over the entire range of posthunt populations. Furthermore, by multiplying the rate at a given posthunt population by the population size, one can derive the average annual number of chronic mortalities if that population size is maintained over time. This allows an estimation of the point at which chronic mortality begins to become more than a trivial, chance event. In this case, 0.05 was chosen as representing more than trivial mortality. The equation also allows an independent estimate of K carrying capacity, since K would lie where the number of chronic mortalities per year equals the number of recruits per year as derived from the recruitment model.

The estimates are shown in relation to the recruitment rate model in figure 7.5. The estimate of the K carrying capacity from this method is 176, as compared to the previously derived 174. The

FIG. 7.5. Recruitment regression as modified by chronic mortality to give a population change model

close agreement is due to the two estimates being opposite sides of the same coin. In the first case, the age structure of living posthunt animals was analyzed to give the maximum survivorship schedule possible, while in the latter, the probabilities of survival in relation to the maximum longevity were analyzed. Again, the internal consistency of the data required that they be in reasonable agreement. Still, to have two independent approaches, with all of the sampling error and complexity involved, give virtually identical results lends credence to the appropriateness of the model.

Since the mean age approach involves subjective judgment of the shape of the survivorship curves and the maximum age approach does not, the latter estimate of 176 was the one retained. Chronic mortality as an important factor begins at approximately MSY, which occurs at a posthunt population of 99, and becomes equal to recruitment at the K carrying capacity of 176. The relationship of chronic mortality to posthunt population size is curvilinear, according to the equation $y = 0.74004 - 0.00047x - 0.00021x^2$. Figure 7.5 shows the relationship of the recruitment curve and the natural mortality curve. Therefore, the population can be modeled through the use of two equations: the linear regression of recruitment up to a posthunt population size of 99, and the curvilinear equation at posthunt populations of 100 and greater.

The relationship between hunting mortality, chronic mortality,

TABLE 7.2. Relationship of Posthunt Population Size and Recruitment Number to Hunting, Chronic, and Prerecruitment Mortality for the George Reserve Deer Population

PoHN *No.*	*Recruit. No.*	*Hunting Mort. No.*	*Hunting Mort. %*	*Chronic Mort. No.*	*Chronic Mort. %*	*Prerecruit. Mort. No.*	*Prerecruit. Mort. %*	*Total Mort. No.*
100	48.68	48.68	100.00	0.00	0.00	0.00	0.00	48.68
110	48.05	47.53	94.12	0.52	1.03	2.45	4.85	50.50
120	46.42	45.45	84.78	0.97	1.81	7.19	13.41	53.61
130	43.78	41.75	74.73	2.03	3.63	12.09	21.64	55.87
140	40.15	36.30	63.76	3.85	6.76	16.78	29.47	56.93
150	35.52	29.99	53.15	6.53	11.57	20.91	37.05	56.43
160	29.89	19.68	36.44	10.21	18.91	24.11	44.65	54.00
170	23.26	8.24	16.75	15.02	30.53	25.93	52.71	49.19
175	19.57	0.00	0.00	19.57	42.76	26.20	57.24	45.77

and prerecruitment mortality is shown in table 7.2 and figure 7.6. At MSY, hunting accounted for virtually all of the mortality. However, with lighter hunting removals, chronic and prerecruitment mortality increase. Prerecruitment mortality increases most rapidly at first, then begins to level off as hunting mortality is decreased and the population approaches K carrying capacity. Chronic mortality increases progressively as hunting mortality is reduced.

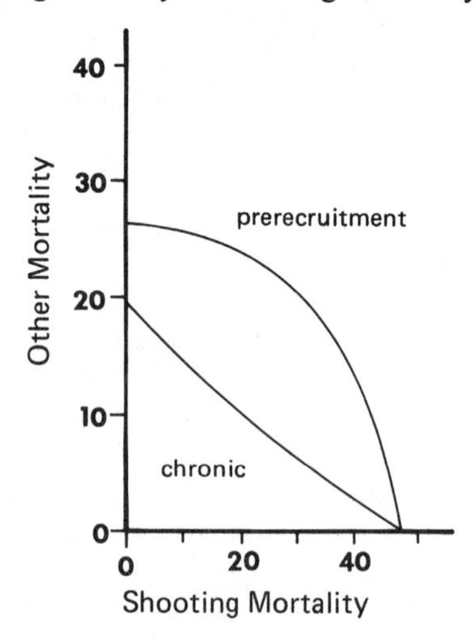

FIG. 7.6. Relationship of shooting mortality to prerecruitment and chronic mortality in George Reserve deer

CHAPTER 8

Population Model Summary

This section will summarize the equations necessary for a deterministic population model for the George Reserve deer population. All the equations have been derived from the empirical data in earlier sections. In addition to summarization, the outcome of this model will be compared to the Pearl-Verhulst logistic equation and expressed as a Ricker-like (1954) stock-recruitment model.

Equations for Population Parameters

The basic datum point for this model is the size of the posthunt population. This value must be specified. Once this value is known, the following population parameters can be estimated by the following equations or procedures.

Sex Composition

1. Y+Ad animals. Multiply posthunt population by 0.5277 to obtain number of females, or 0.4773 to obtain number of males (assumes unselective hunting).
2. Recruits or embryos. Multiply number of recruits or embryos by 0.4659 to get average number of females, or 0.5341 to get average number of males (p. 71); or use regression $y = 30.1181 + 0.2081x$, where x is posthunt population size and y is the percentage of offspring which is male (p. 68).

Age Distribution of Females

3. Number of fawn females. If the posthunt population number of females is *less than 55*, fawn females (y) can be obtained by the equation $y = 0.892 + 0.3543x$, where x is the posthunt female population (p. 59).
4. If the posthunt number of females is *55 or greater*, the equation is $y = 13.7265 + 1.411x - 0.0097x^2$ (p. 60).

Embryo Rates

5. Embryo rate for fawn females. This rate can be obtained by the equation $y = 1.8681 - 0.0917x$, where y is the embryos per fawn female and x is the posthunt number of fawn females (p. 62). Number of posthunt fawn females is obtained by equation 3 or 4.
6. Embryo rate for Y+Ad females. This rate is obtained by $y = 1.9461 - 0.0175x$, where y is the embryo rate per female and x is the number of posthunt Y+Ad females (p. 62) and where the latter is obtained by subtracting the number of fawn females (equation 3 or 4) from the total of posthunt females (equation 1).

Embryo Numbers

7. Embryo number of fawn females. This number is obtained by multiplying the embryo rate (equation 5) by the posthunt number of fawn females.
8. Embryo number of Y+Ad females. This number is obtained by multiplying the embryo rate (equation 6) by the posthunt number of Y+Ad females. The total number of embryos is obtained by adding the numbers given by equations 7 and 8.

Recruitment Rate

9. Recruitment rate is obtained by the equation $y = 0.9868 - 0.005x$, where y is the recruitment rate and x the total posthunt population (p. 99).

Recruitment Number

10. Recruitment number is obtained by multiplying the recruitment rate by the total number of animals in the posthunt population.

Change in Population Growth Rate

11. If the posthunt total population is *99 or less*, the recruitment rate (equation 9) is the population growth rate (p. 99).
12. If the total posthunt is *100 or greater*, the equation is $y = 0.7400 - 0.00047x - 0.000021x^2$, where y is the rate of change and x is the total posthunt population (p. 113). Note that the rate becomes negative with posthunt populations of 176 or greater.

Change in Population Number

13. The change in population number is obtained by multiply-

ing the population change rate (equation 11 or 12) by the size of the total posthunt population.

Prerecruitment Mortality

14. Prerecruitment mortality, i.e., the number of embryos failing to survive to recruitment age (approximately six months of age), is obtained by subtracting the number of recruits (equation 10) from the total number of embryos (equations 7 and 8).

Chronic Mortality

15. Number of animals dying of chronic causes is obtained by subtracting the population change in numbers (equation 13) from the number of recruits (equation 10).

Equilibrium Values

This series of equations was used to generate population estimates at equilibrium for ten-animal increments of posthunt total population over the range of possible values (table 8.1). Equilibrium refers to a set of parameter estimates from the model that yield a steady state of given population size. Since mean values are used in the equation, the model is deterministic and generates a set of fixed values, given a starting value of the posthunt population. In nature, of course, there is considerable variation. Nevertheless, there is a functional relationship between the prediction of this deterministic model and the George Reserve deer herd. Assume that the posthunt population is stabilized at, for example, 120 animals by shooting, and that each year the number of animals necessary to return the population to 120 head would be removed by hunting. If we did this for ten years, even though the parameters varied considerably between given years, the average values for the ten-year period would approximate those given for the 120 posthunt population in table 8.1. The same would be true of any other posthunt population selected.

The boxed columns in the table represent important points in the posthunt population. The MSY point is 99 posthunt population, while K carrying capacity is 176. Posthunt populations greater than 176 are inherently unstable, i.e., an equilibrium does not exist. They are included, however, to indicate the posthunt populations at which zero recruitment (198) and zero embryos (219) would occur, and to illustrate the magnitude of posthunt population sizes over which these parameters operate.

TABLE 8.1. Equilibrium Values for the George Reserve Deer Population if Maintained at a Given Posthunt Population

	Posthunt Population									*MSY*	
	10	20	30	40	50	60	70	80	90	99	100
Total No. ♂ ♂	4.77	9.55	14.32	19.09	23.87	28.64	33.41	38.18	42.96	47.25	47.73
Total No. ♀ ♀	5.23	10.45	15.68	20.91	26.13	31.36	36.59	41.82	47.04	51.75	52.27
No. Fawn ♀ ♀	2.74	4.59	6.45	8.30	10.15	12.00	13.86	15.71	17.56	19.23	19.30
No. Y+Ad ♀ ♀	2.49	5.86	9.23	12.61	15.98	19.36	22.73	26.11	29.48	32.52	32.97
Total No. Embryos	9.29	17.74	25.17	31.59	36.95	41.31	44.63	46.93	48.19	48.44	48.71
No. ♂ ♂ Emb.	4.96	9.48	13.45	16.87	19.74	22.07	23.84	25.06	25.74	25.87	26.01
No. ♀ ♀ Emb.	4.33	8.26	11.72	14.72	17.21	19.24	20.79	21.87	22.45	22.57	22.70
Emb. by Fawn ♀ ♀	4.43	6.64	8.23	9.19	9.51	9.21	8.28	6.72	4.53	2.01	1.90
Emb. by Y+Ad ♀ ♀	4.86	11.10	16.94	22.40	27.44	32.10	36.35	40.21	43.66	46.43	46.81
Total Recruitment	9.37	17.74	25.10	31.47	36.84	41.21	44.58	46.94	48.31	48.69	48.68
No. ♂ ♂ Recruit.	5.00	9.47	13.41	16.81	19.68	22.01	23.81	25.07	25.80	26.00	26.00
No. ♀ ♀ Recruit.	4.37	8.27	11.69	14.66	17.16	19.20	20.77	21.87	22.51	22.69	22.69
Total Mortality	9.37	17.74	25.10	31.47	36.84	41.21	44.58	46.94	48.31	48.69	48.68
Hunting Mort.	9.37	17.74	25.10	31.47	36.84	41.21	44.58	46.94	48.31	48.69	48.68
Chronic Mort.	0.00	0.00	0.00	0.00	0.00	0.00	0.00	0.00	0.00	0.00	0.00
Prerecruit. Mort.	0.00	0.00	0.00	0.00	0.00	0.00	0.00	0.00	0.00	0.00	0.00

	Posthunt Population						*Zero Pop. Change*			*Zero Recruit.*			*Zero Emb.*
	110	120	130	140	150	160	170	176	180	190	198	200	210	219
Total No. ♂ ♂	52.50	57.28	62.05	66.82	71.60	76.37	81.14	84.00	85.91	90.69	94.51	95.46	100.23	104.53
Total No. ♀ ♀	57.50	62.72	67.95	73.18	78.41	83.63	88.86	92.00	94.09	99.31	103.49	104.54	109.77	114.47
No. Fawn ♀ ♀	19.67	19.51	18.82	17.59	15.84	13.55	10.73	8.79	7.37	3.49	0.00	0.00	0.00	0.00
No. Y+Ad ♀ ♀	37.83	43.21	49.13	55.59	62.57	70.08	78.13	83.21	86.72	95.82	103.49	104.54	109.77	114.47
Total No. Embryos	50.50	53.61	55.87	56.93	56.43	54.00	49.19	45.77	41.56	30.67	19.23	17.51	8.33	0.00
No. ♂ ♂ Emb.	26.97	28.63	29.84	30.41	30.14	28.84	26.27	24.45	22.20	16.38	10.27	9.35	4.45	0.00
No. ♀ ♀ Emb.	23.53	24.98	26.03	26.52	26.29	25.16	22.92	21.32	19.36	14.29	8.96	8.16	3.88	0.00
Emb. by Fawn ♀ ♀	0.00	0.00	0.00	0.00	0.00	0.00	0.00	0.00	0.00	0.00	0.00	0.00	0.00	0.00
Emb. by Y+Ad ♀ ♀	50.50	53.61	55.87	56.93	56.43	54.00	49.19	45.77	41.56	30.67	19.23	17.51	8.33	0.00
Total Recruitment	48.05	46.42	43.78	40.15	35.52	29.89	23.26	18.80	15.62	6.99	0.00	0.00	0.00	0.00
No. ♂ ♂ Recruit.	25.66	24.79	23.38	21.44	18.97	15.96	12.42	10.04	8.34	3.73	0.00	0.00	0.00	0.00
No. ♀ ♀ Recruit.	22.39	21.63	20.40	18.71	16.55	13.93	10.84	8.76	7.28	3.26	0.00	0.00	0.00	0.00
Total Mortality	50.50	53.61	55.87	56.93	56.43	54.00	49.19	44.99						
Hunting Mort.	47.53	45.45	41.75	36.30	29.99	19.68	8.24	0.32	Inherently unstable					
Chronic Mort.	0.52	0.97	2.03	3.85	6.53	10.21	15.02	18.48	Population will decline					
Prerecruit. Mort.	2.45	7.19	12.09	16.78	20.91	24.11	25.93	26.19						

Hunting is the only variable, other than posthunt population size, that is external to the model. Perhaps some explanation of the hunting mortality in table 8.1 is required. From the MSY posthunt population level and below, hunting is the only source of mortality in the population. Therefore, if the posthunt population is to remain in equilibrium, the average hunting kill must just match the average recruitment. If recruitment exceeds kill, the posthunt population size will increase; but if recruitment is less than the kill, the posthunt population will decline. At slightly above the MSY posthunt population, chronic mortality begins. This mortality is based upon an estimate of the probability that some animals will die of chronic causes even though a fairly high number are being shot. Thus, at posthunt populations in excess of 100 individuals, the sum of the hunting and chronic mortalities must, on average, equal the recruitment if equilibrium is to be maintained. If the combined mortality is less than the recruitment, the posthunt population will increase; if it is greater, the population will decrease. If hunting mortality at a given equilibrium is increased, the posthunt population will decline to the point at which recruitment comes into balance; or, if the kill per year exceeds the maximum recruitment possible (48.69 animals at a posthunt population of 99), the population will decline to extinction. If hunting mortality is decreased, the posthunt population will grow to the level at which the combined hunting and chronic mortality equal recruitment. The impact of varying the hunting kill will be presented in greater detail in the following chapter.

Comparison to the Logistic Equation

The logistic equation as a population model has had its critics and its defenders, but despite its critics, no mathematical formulation has had a greater impact upon the field of ecology (see May, 1976). In many instances, it is treated as a "law" of population growth. Moreover, the logistic equation has been the starting point for most of the mathematical treatments of interspecific competition and predator-prey interactions that have had pronounced influence upon the development of niche theory and food-chain resources. The number of models and simulations, the sheer mass of literature in ecology which rests on the logistic growth equation, staggers the mind.

A comparison of the outcome of this empirically derived deer

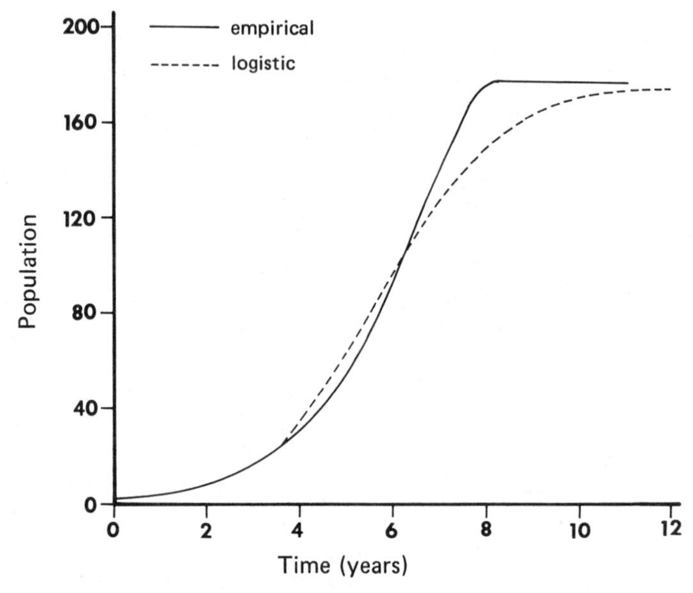

FIG. 8.1. Predicted population growth curve from the George Reserve model as compared to the logistic equation

population model to a population growth curve derived from a logistic equation can be obtained by generating a population growth curve from the model assuming a small initial population. If one starts with an initial population of a single pair of adult deer and follows a process in which each year (t) is an iteration of the deterministic population model, a population growth curve can be obtained (fig. 8.1; table 8.2). Superficially it resembles a logistic equation growth curve, but recall that it is derived in a fundamentally different way.

The logistic equation

$$N_t = \frac{K}{1 + e^{a - r_{max} t}}$$

can be used to generate a population growth curve for comparison with the empirical model growth curve. The estimate of K carrying capacity is 176. The value of r_{max} can be estimated from empirical data by the method of Pearl (1930), in which $\log_e$ K−N/N is regressed on time; a is the y intercept of zero x, and r_{max} is the slope. Using the population sizes on time of the empirical model, esti-

TABLE 8.2. Population Growth Derived from the George Reserve Deterministic Model (assuming no hunting mortality)

Time	*Pop. No.*	*Recruit.*	*Pop. Change*	*Chronic Mort.*	*Prerecruit. Mort.*
0	2	1.95	1.95	0	0
1	3.95	3.82	3.82	0	0
2	7.78	7.37	7.37	0	0
3	15.15	13.80	13.80	0	0
4	28.95	23.38	23.38	0	0
5	53.33	38.40	38.40	0	0
6	91.73	48.45	48.45	0	0
7	140.18	40.08	36.20	3.88	16.86
8	176.38	18.51	−0.20	18.71	26.18
9	176.17	18.66	0.08	18.58	26.19
10	176.25	18.60	−0.03	18.63	26.19
11	176.22	18.63	0.01	18.61	26.19
12	176.23	18.62	−0.01	18.62	26.19
13	176.23	18.62	0.00	18.62	26.19

mates of $a = 4.6507$ and $r_{max} = 0.8043$ were derived. Of course, realized rates of r usually are substantially lower. For example, the early growth rate of the George Reserve deer population following introduction (which is a classical example of rapid growth) was $r = 0.5678$.

Using these values of a and r_{max} in the logistic equation gives the growth curve shown in figure 8.1. The comparison with the empirical model is not good. The logistic equation overestimates population size during the early years and underestimates it in later years. While these deviations seem minor on gross inspection, more careful examination will show an important discrepency. The equation fails to express the abrupt leveling off of population growth as K carrying capacity is approached. Recall that the empirical model, as given at this stage, is "controlled" in the sense that time lags are not permitted. But this growth pattern emphasizes the great likelihood of overshoot of K, an inherent characteristic of the actual nature of the animal under consideration.

Thus, the estimate of time required to achieve subsistence population is greatly overestimated by the logistic equation. Rounding population size estimates to the nearest whole animal, the logistic equation gives a population of 176 in fourteen years, while the empirical model reaches 176 in eight years—a substantial difference. In conclusion, the logistic equation does not give a particularly precise picture of how the population appears to behave based on empirical results. The difference in the two models traces to the fact that the logistic equation assumes that r decreases linearly with density (Wilson and Bossert, 1971; Krebs, 1972), while in the empirical model, the relationship becomes curvilinear as the population size becomes large. The essential difference between the two models is that the relationship between reproduction and mortality is determined in the empirical model, rather than assumed as in the logistic model.

May (1976), in his synoptic treatment of single population models, noted the difficulty created by populations with distinct but overlapping age classes. To this must be added the further ramifications of qualitative differences within and between age classes, such as sex, physiological condition, social status, and others. Furthermore, the logistic equation achieves population change through an artificial process relative to the real function in nature. In the equation, r_{max} is a constant, while in nature it is a variable. The environmental resistance factor (K−N/K) represents remaining food and is necessary to adjust r to be variable. In nature, population growth is not dependent upon the amount of food remaining but rather the amount of resources consumed, digested, and converted into new tissues and new individuals. The logistic equation treats food as a constant, not a dynamic entity (Caughley, 1976) and eliminates the possibility of threshold effects. In short, the logistic equation produces growth by a process which is not analogous to the process in nature.

It is my opinion (or perhaps, bias) that a good population model should mimic the process in nature. A model that mimics nature is less likely to be fooled by the extremes of nature. Furthermore, if the model proves to be inadequate for the purpose, that natural process which is important but unaccounted for can be added to the model for correction. Artificial process models give answers for the wrong reasons. So long as they give the right answers, one is not likely to worry that the reasons are wrong, but if they give wrong answers, how are they to be corrected? One can add another incorrect reason, a move that will correct that answer but makes the model even more artificial and even less likely to track correctly the vagaries of nature. I agree with Caughley and Birch (1971) that given no information on a real world problem, the logistic growth curve is at least a useful starting point. But in a specific instance, even a small amount of empirical data is more useful.

The fact that the growth of the George Reserve deer population does not fit the logistic equation is not sufficient evidence for discarding it as a model, but it should give the most ardent supporters of the equation reason for caution. Given a value of K, a starting point, and relatively small scales on the x- and y-axes, there are only a limited number of S-shaped curves which are possible. It is inevitable that the logistic growth curve will approximate some of them. Certainly the logistic model has and will continue to have considerable heuristic value, but examining what we mean by a "general" model is perhaps long overdue. We may have fallen into thinking it is all-encompassing, rather than a crude first attempt. The distinction is important, given the continuing efforts to wring more and more detail from the natural world based upon the logistic model. It may be hazardous to have so much of ecological theory resting upon so unfirm a foundation.

Expressed as a Ricker-like Stock-Recruitment Model

The empirical model can be expressed as a Ricker-like (1954) model by plotting recruitment of the population at a given size above the 45-degree slope, which would pertain if no recruitment were present (fig. 8.2). If the recruitment line thus plotted is extended by assuming the regression of figure 7.5 crosses the x-axis giving negative values of recruitment, one gets a Ricker-like curve that will account for overshoots and subsequent return to K carrying capacity.

Figure 8.2 demonstrates several important characteristics of the George Reserve deer population. First, assuming that time lags are absent, there is a fairly strong equilibrium tendency at K. As pointed out by Moran (1950) and Ricker, if the recruitment curve crosses the forty-five-degree line at a shallow angle, the equilibrium tendency is strong. The relatively shallow angle of the figure is what one would expect from a K-selected species such as deer. The tendency to achieve equilibrium decreases as the angle at which the recruitment curve crosses the forty-five-degree line increases, until at a perpendicular or greater angle, equilibrium is achieved only if the posthunt population happens to fall at K. The perpendicular crossing can be compared to the recruitment rate regression (fig. 6.5) to gain an understanding of the nature of the equilibrium. Of the constants in the regression equation $y = a + bx$, a (the value of y at $x = 0$) is the critical value. Given the value of a, b (the slope of the regression) can assume any negative value,

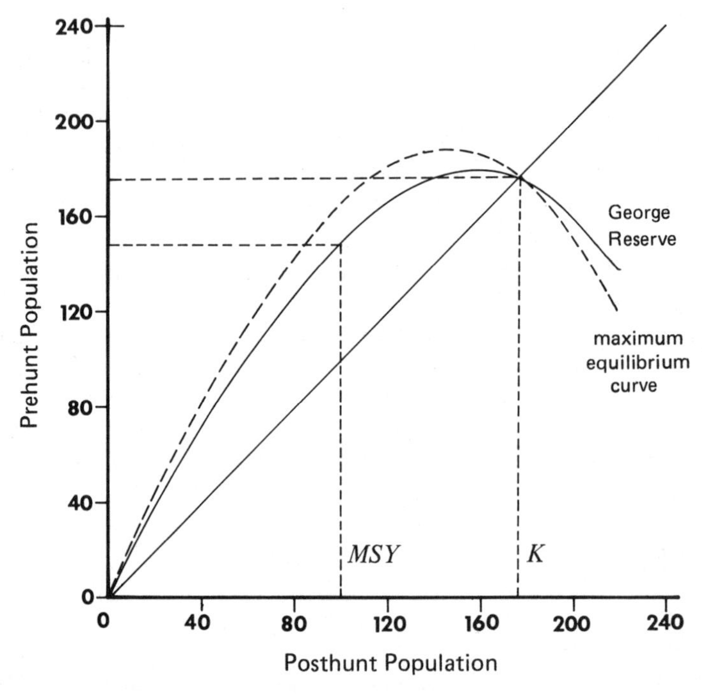

FIG. 8.2. Ricker-like stock and recruitment model of the George Reserve deer population

and the slope at which the recruitment curve crosses the forty-five-degree line in a Ricker-like model will not change, although the point at which it intercepts the x-axis will increase as the slope decreases. Thus, *a* is the critical constant in determining whether the model will show an equilibrium tendency. Equilibrium tendency declines as the value of *a* increases. The value of *a* at the model's perpendicular crossing of a linear population change curve has a steeper angle of crossing, and correcting for this variable (assuming a longevity similar to deer) lowers *a* at maximum equilibrium to about 1.26. A longer life span would increase this value slightly, while a shorter life span would decrease it. Longevity in this calculation is not a very sensitive parameter. Therefore, *a* of maximum equilibrium will not vary greatly from 1.26. If a recruitment rate regression has a value of *a* approximately equal or less than this, it will achieve equilibrium. If the value is substantially less as in figure 6.5 ($a = 0.9868$), a strong equilibrium tendency will be present, while as the value of *a* approaches 1.26, gradually dampening oscillations will eventually lead to equilibrium. If a exceeds 1.26, equilibrium will never be achieved (unless the value of the crossing point in the Ricker model happens to occur by chance; this chance effect is a quirk of the model, and is unrealistic biologically). As the value of a increases, the tendency of the population to go to extinction also increases.

Note that the value of a in the recruitment rate regression is derived by using the total population. If using females only, the value of a increases substantially. Thus, if only females are used on the x-axis, female recruits should be used on the y-axis to calculate a and determine if the population shows equilibrium tendencies. Similarly, it is typical in life table analyses to use females only; as long as fecundity rates are for female offspring only, derivation of r from life table analysis should give comparable results. Derivation of r_{max} from a life table, of course, requires that the data come from a population growing at an unimpeded rate in an optimum environment. Such situations are seldom observed for vertebrates under natural conditions. Most are related to introductions or severe man-caused manipulations, although naturally occuring introductions or population recovery following catastrophic events do occur infrequently.

The same phenomenon relative to the net reproductive rate (R_0) of the life table was explored by Maynard-Smith (1968). Assuming that R_0 shows a linear decline with increasing population size, the equilibrium tendency can be judged by the equation $L = bN_k$, where b is the slope of the regression of R_0 on population size, and N is the population size at K equilibrium (notation changes have been made to avoid confusion with other terms used here). Values of L between zero and one lead to stable equilibrium at K; values of one to two lead to gradually dampening oscillations towards K; and values of two or greater lead to increasing oscillations resulting in eventual extinction. Indeed, $L = bN_k$ is the formula for deriving a of the regression equation, so if the relationship is linear, $L = a$. Recruitment rate and R_0 are measuring the same population phenomenon. Equilibrium R_0 at any given population size can be derived from table 8.1. R_0 does not show a linear relationship to population size (see fig. 7.5), but the average slope of the curvilinear relationship is 0.0054. Thus, for the George Reserve deer population $L = 0.0054\ (176) = 0.9504$. If R_0 were linear, the deer population should achieve a stable equilibrium without overshoot. But the actual slope at K of the curvilinear regression is 0.006, giving an L of 1.06. Thus, a slight tendency to overshoot K and quickly dampen to K is suggested by the outcome, a seemingly correct response for the George Reserve deer herd.

These values of L can be related to the angle of crossing of the forty-five-degree line by the recruitment curve in the Ricker-like model. L values of two or greater exceed the perpendicular crossing and lead to extinction. L values of one to two are cases where the angle of crossing is less than the perpendicular but where the highest point of the recruitment curve exceeds a horizontal line passing through the crossing point. Such a case will result in overshoot with gradual dampening oscillations leading to K. The George Reserve deer herd shows this situation, for the highest point of its recruitment curve slightly exceeds a horizontal line through the crossing point (fig. 8.2). L values of one or less are equivalent to the recruitment curve not exceeding the horizontal line through the crossing point, and these lead directly to equilibrium at K carrying capacity. Caughley (1976) presents the same relationship relative to the logistic equation with a time lag.

A second point illustrated by figure 8.2 is the relationship of total standing crop to population density: at K carrying capacity there is a peak population standing crop of 176 animals; however, at MSY it is only 147.7. Manipulating the population from the K carrying capacity in the direction of MSY inevitably reduces the size of the peak population standing crop in this species. Even the maximum equilibrium curve shows some decrease of standing crop of animals at MSY over that of K. Thus, populations which have a greater standing crop at MSY than at K will, by definition, not achieve equilibrium at K. The lower the r_{max} of the population, the greater the decrease of the peak population standing crop by MSY below that at K. Since white-tailed deer have the highest r_{max} of any North American ungulate, the same would be true of all North American ungulate populations. Climax species that have curves approximately one-half as high above the forty-five-degree slope as the George Reserve deer would have relatively greater discrepancies in standing crop between MSY and K.

The behavior of the empirical model can be explained in terms of the Ricker-like model, since both posthunt population (prerecruitment) and prehunt (postrecruitment) are shown in the same figure. The forty-five-degree line in figure 8.2 illustrates a situation where no recruitment occurred (note that the scale of values on both the y- and x-axes are the same). Therefore, if a given posthunt population showed no net change, the posthunt population in the following year would be the same. One would begin at the point of a given posthunt population at a given time (t_0), read up to the forty-five-degree line, and over to the y-axis to the same posthunt population size. Completion of one cycle (in this case a year) involves returning to the x-axis along the hypotenuse of a right triangle, with the triangle's apex located at intersections of the x-and y-axes. Thus, at time t_1, you would be back at the beginning posthunt population level. As long as there was zero recruitment, one would repeat each cycle along the same pathway, adding one to the time for each cycle.

Zero recruitment is impossible over any length of time without the population declining. The curved line in figure 8.2 represents the recruitment of new individuals to the population. In essence, it is the plotting of the parabola from figure 6.6 above the forty-five-degree line in figure 8.2. Recall that from zero posthunt population up to ninety-nine, the recruitment rate regression is used because no chronic mortality occurs. However, with a posthunt population of one hundred or greater, the value of the parabola is the number of recruits minus the chronic mortality, which actually gives a net population change.

Including recruitment allows figure 8.2 to be used as a predictive model. The process of reading from a given posthunt population is the same, but the recruitment curve is the point used in arriving at the y-axis, rather than the forty-five-degree line. Beginning at a given posthunt population size on the x-axis, for example forty, and reading to the y-axis from the recruitment curve we get 71.5. Assuming that no animals are removed, the population at time $t+1$ will be 71.5 and this will be the new posthunt population. Repeating the cycle gives 116.5. Again, if no animals are removed, 116.5 is the new posthunt population at the end of $t+2$. Continuing the cycles will lead to equilibrium at the K carrying capacity of 176. Excluding the problem of reading the graph accurately, the Ricker-like model gives the same predictions as the equations derived earlier.

Although the Ricker-like model can be used to make predictions, computer programs are far more convenient. They are rapid, precise, and not prone to error. In the following chapters, the model equations are examined in detail in terms of various management computer simulations.

CHAPTER 9

Yield and Stability: Model Predictions

Yield is usually defined as the weight or number of animals which are taken from a population by man (Ricker, 1975). Since numbers are the main focus of this study, *yield* as used here will refer to the number of individuals taken by man.

A yield is possible at any population size above the minimum number required for effective breeding. In white-tailed deer, it is one male and one female. On the George Reserve, yield is equivalent to the number of animals added to the population at any given population size (hunting mortality, table 8.1). For example, at a posthunt population of 10, an average yield of 9.37 animals can be taken; at 30, 25.10; and at 90, 48.31. Maximum sustainable yield (MSY) occurs at a posthunt population of 99, at which the yield is 48.69.

Equilibrium yields at populations above MSY level are also indicated in table 8.1. This is the number that must be harvested by man in order to stabilize the population at a given posthunt number. Note that, on average, any removal of up to MSY can be taken from a population greater than that yielding MSY, but if such a yield is greater than the equilibrium yield, the population will be reduced to the point where the actual yield is equal to the equilibrium yield. Because of variation in population performance between years, the maximum fixed removal which does not lead to extinction will be somewhat lower than MSY.

The following specific definitions of yield can be given, and are based on the assumption that there are no changes in existing environmental conditions: *Sustained yield* (SY) is the average yield that can be taken from a population which results in a stable equilibrium of the posthunt population at a given size (Caughley, 1976). With fluctuating recruitment, fewer animals would have to be taken in some years than in others. *Maximum sustained yield* (MSY) is the maximum average number of animals that can be removed from a population without leading to extinction (Ricker, 1975). With fluctuating recruitment, fewer animals would have to be taken in some years than in others. *Fixed removal yield* (FRY) is the maximum fixed number of animals that can be removed from a population with fluctuating recruitment without driving the population to extinction. *Optimum sustained yield* (OSY) is that yield which the population can sustain and that maximizes human benefits. Such benefits may be economic return (Caughley, 1976) or esthetic values, such as hunter satisfaction. Since hunting was the controlled variable on the George Reserve and is a major variable within the control of wildlife management agencies, the impact of a given harvest on the population is of prime interest. This impact can be studied through simulations of the effect of size of kill on the George Reserve deer population model.

Deterministic Model Predictions

The model derived thus far is simplified in two respects: it does not contain a stochastic element, and time-lag effects have been omitted. Nevertheless, it is desirable to observe the characteristics of the simplified model to get the basic responses of the population before dealing with the complexities of a more realistic model.

To a considerable extent, the George Reserve deer population has been replacing removals by an equal recruitment. For example, a decision to hold the posthunt population at 120 would require, on average, the removal of 46.42 animals per year (table 8.1). The process has been to count the herd and remove the excess above 120. The same effect could have been achieved by removing 46.42 animals per year, in which case the prehunt population would have averaged 120 over several years. This "equilibrium" achieving effect occurs on the right side of the parabola in figure 6.6. Note that assuming no variation, a similar equilibrium could be achieved on the left side where a removal of 46 is achieved at a prehunt population of 76. Similarly, all other removals of fewer than 49 deer per year have equilibrium points on both sides of the parabola.

Fractions of animals are reasonable when one is dealing in long-term averages. However, in a given year the population and removals are made in whole animals, for one cannot remove 46.42 animals. Therefore, model predictions are rounded to the nearest whole animal each "year" or iteration hereafter. Note that this is the way the deer population actually operates.

If one is at an equilibrium on the left side of the parabola and the number removed per year is increased, the population will be driven to extinction; if the number removed per year is deceased, the population will increase to the equilibrium for the new number removed on the right-hand side of the parabola. An example is shown using one set of values in figure 9.1. Changing the values changes the numbers obtained, but not the basic response. On the left side of the parabola (fig. 6.6), exceeding the current recruitment results in a deceleration towards extinction. Removing less than the current recruitment results in an accelerating population growth to the equilibrium point on the right side of the parabola. Therefore, stabilization on the left side of the parabola can only be achieved by an exact removal of the recruitment at that particular population density. Population responses on the right side of the parabola can be predicted by assuming a given population size and a fixed removal. In figure 9.1 the population is allowed to increase to the K carrying capacity (176) and then treated with fixed removals of 10, 20, 30, 40, 49, (MSY), 50, and 60 per year. Removals of 10 through 40 per year are easily balanced by the potential for recruitment. There is virtually no time lag between the beginning of the annual removal and achievement of equilibrium. For removals up through 30 per year, no reduction of peak population size occurs. Indeed, the population increases slightly above that at K carrying capacity (10 removals, 179; 20, 180; 30, 179). At removals of 37 or more per year, peak population at equilibrium begins to decline from that at K carrying capacity.

An average kill of 49.69 per year is the maximum that the population can support (fig. 6.6). Under a removal of 40 per year, the population first decreased rapidly and then more gradually over time until equilibrium was reached (fig. 9.1). Removal of more than an average of 49.69 per year exceeds the maximum recruitment and extinction is inevitable. Removal of 50 per year is only a bit greater than 49.69 and the population decreases rapidly at first, then begins to approach equilibrium. However, as peak recruitment numbers are passed (moving from right to left, fig. 6.6), a deceleration to extinction occurs. The greater the kill exceeds the MSY, the more rapidly extinction occurs. Again, this set of values is for illustration. Any values can be used, but the basic response will remain the same. Furthermore, any combination of values can be used to observe the behavior of the model. One need not follow the pattern

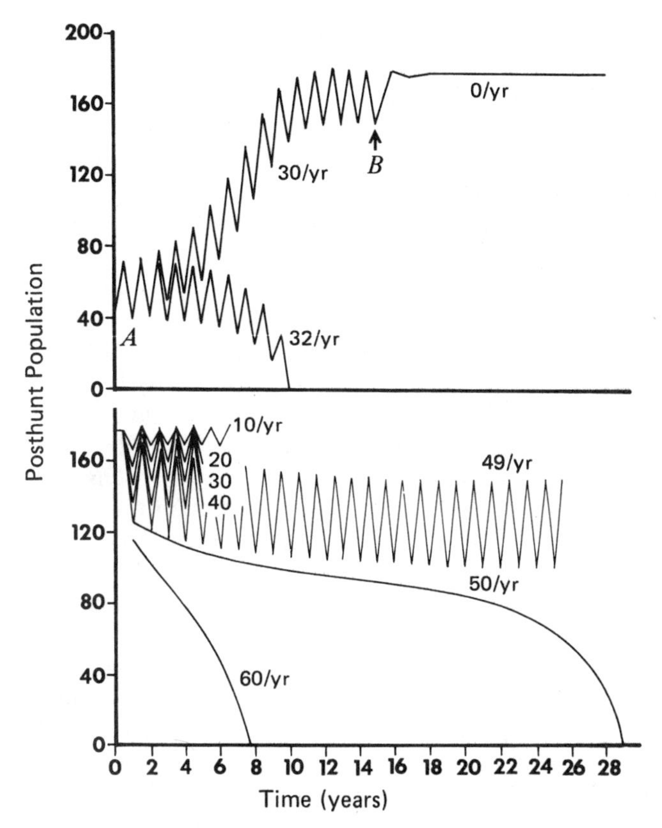

FIG. 9.1. Outcomes of fixed removals. Above, starting from a posthunt population of 40 animals at *A*, removal of 32 per year leads to extinction, while 30 per year leads to equilibrium. Stopping all removals at *B* leads to a fixed equilibrium. Below, starting from K carrying capacity (176), removals of less than 50 per year leads to equilibrium while greater than 50 per year leads to extinction. Only posthunt population is plotted for kills of 50 and 60 per year.

of a fixed removal per year. There are an infinite number of combinations that can be followed.

Stochastic Population Change Model and Fixed Kill

Having determined how the model behaves in the simple form, we can now study the influence of stochastic processes—that is, variation about the average in given years.

With population change varying stochastically, killing the MSY (49 per year) will lead inevitably to extinction. Sooner or later random variation will result in a net gain lower than 49, and the kill will move the population to the left-hand side of the recruitment number parabola (fig. 6.6). If a subsequent, extraordinarily high net gain does not occur the following year to bring the population numbers to the top of the parabola, extinction will be inevitable. Maximizing the kill of a mean of 49 would require knowing the exact number of individuals gained in a given year and removing this number in the kill. Since the number is virtually never known with a wild ungulate, even in the George Reserve, the kill must be constrained to a lower level where stochastic variation in net gain is low enough that extinction is highly improbable.

The computer program was altered to simulate population responses with stochastic variation in population change and the kill fixed to a given number. Stochastic variation in the regression rate of population change on posthunt population was introduced by using random values based on the standard deviation of the recruitment regression equation for total population (chap. 6). The adjusted recruitment rate was multiplied by the posthunt population to give the stochastically adjusted number of young recruited. Net population change was added to or subtracted from the posthunt population, yielding a new prehunt population from which the fixed kill was removed and the process repeated. Fixed kills were used first to determine the influence of stochastic variation in population change without the influence of variation in the kill. The starting point for prehunt population used in the simulations was K carrying capacity (176). The program contained three nested loops to perform iterations. Simulations at a given kill were repeated until the population went to extinction or reached 100 years without extinction. At the end of a given run, the prehunt population was returned to 176 and the process repeated with the same kill until twenty cases of the given kill were obtained. When those cases were completed, the kill was increased by one, and twenty cases at the new kill accumulated. Thus, the fixed kill was incremented by one until the MSY (49) was exceeded. No extinctions in twenty cases of 100 years each means the simulated population went 2,000 years without an extinction. This is not precisely correct, since the population was returned to K carrying capacity (176) at the end of each 100 years, and a few years may be required to achieve approximate equilibrium. Nevertheless, twenty cases without extinction was considered an acceptable risk of extinction.

TABLE 9.1. Outcomes of Simulations of Stochastic Population Change

	Stochastic Population Change, Fixed Kill		*Stochastic Population Change, Stochastic Kill*	
Kill/yr.	*No. Extinctions*	$\bar{x}$ *Extinction year*	*No. Extinctions*	$\bar{x}$ *Extinction year*
40	0	—	0	—
41	0	—	0	—
42	0	—	0	—
43	0	—	0	—
44	0	—	3	59.3
45	0	—	6	50.8
46	0	—	14	47.1
47	1	65.0	19	36.6
48	12	65.0	18	35.7
49	16	52.5	20	30.4
50	20	26.3	20	20.6

Note that in neither these simulations nor those to follow is any factor included that relates to density-dependent influences upon obtaining the kill. It is reasonable to assume that as deer become less and less abundant in nature, they would be more and more difficult to hunt. This would function as another force working against population extinction. The models ignore this variable and, therefore, overemphasize the possibility of extinction. Estimates of the safe level of kill from the model should be on the conservative side.

Table 9.1 gives the results of the population simulations. Fixed kills up through forty-six animals per year gave no extinctions. A kill of forty-seven gave one extinction, forty-eight gave twelve extinctions, and forty-nine gave sixteen extinctions out of twenty cases. In all cases, a kill of fifty led to extinction. The results demonstrate the remarkable equilibrium tendencies of the system, even including the independent, uncontrolled variation due to unidentified factors influencing recruitment. The difference of three animals between FRY and MSY suggests that approximately 94 percent of MSY can be killed safely. Intuitively, a much lower number was expected.

Some typical runs for twenty-five years are plotted in figure 9.2. Note that the lower kills had little impact on the background fluctuations in the population and, in fact, tended to increase the fluctuations slightly. As kills increased, the stabilizing effect be-

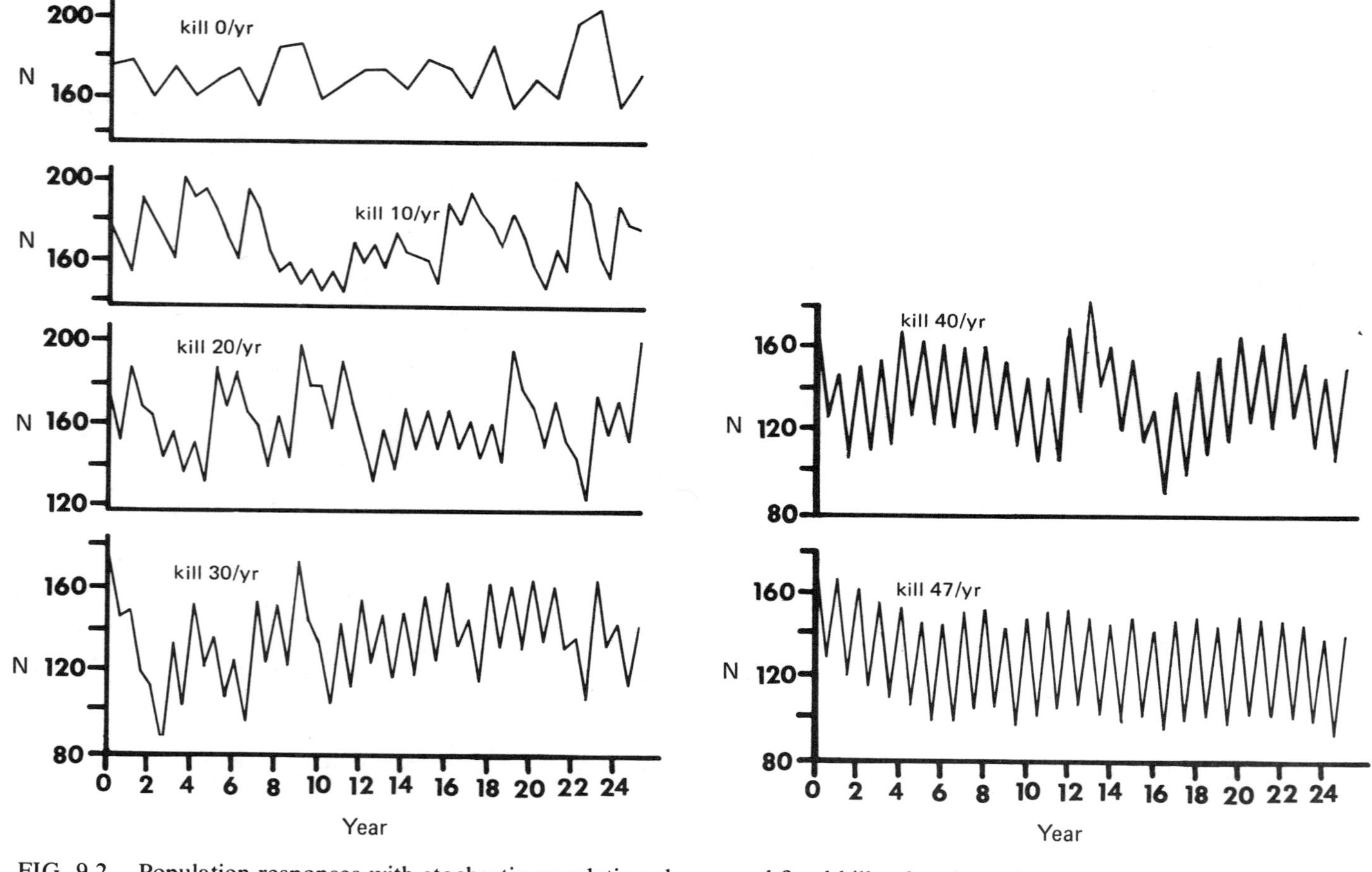

FIG. 9.2. Population responses with stochastic population change and fixed kills of various sizes

came increasingly apparent. A rough measure of the stabilizing effect on harvesting can be obtained by calculating the standard deviation of either the prehunt or posthunt population (because the kill is fixed, the outcome will be the same), since the population is fluctuating randomly about a mean value. The first five years of each simulation in the figure are deleted since the population was always started at 176 and a time lag was present in reaching approximate equilibrium for the higher kills. Thus, a sample size of twenty years was used. The standard deviations were: 0 kill/year, 13.32; 10/year, 16.45; 20/year, 15.82; 30/year, 12.23; 40/year, 11.75; and 47/year, 3.91.

Deterministic Population Model and Stochastic Kill

On the George Reserve it is quite possible to achieve a fixed kill per year. However, in most situations, a fixed kill can be only approximated. It is instructive to assume constraints of management similar to those confronted by a state wildlife management agency to test the response of the empirical model, given that the desired removal is difficult to achieve and actual removal varies substantially.

In most states, the deer kill in the general hunting season is usually moderate, the result from a trade-off between what is known about the capacity of the population to support a high kill and public opposition to shooting antlerless animals (Longhurst, 1957). It is within the legal authority of state wildlife agencies to more closely approximate a fixed kill by setting open-ended seasons. From a given starting date the kill is monitored, and as the desired kill is achieved, the season is closed. This kind of season has been conducted only on an experimental basis. Because of the complexity in making such a program understood, monitoring the kill, communicating the closing date of the season, etc., the approach has not become general practice. Moreover, the rationale upon which a fixed kill might have been based was not so apparent as to counterbalance the difficulties of conducting such a season. Even if the problems could be overcome, the public's acceptance of such an approach is unknown. Public opinion, correct or not, is a major constraint on management programs.

Typically, wildlife officials have established hunting programs based upon a given length of season, times of day when hunting was legal, bag limits with sex or age restrictions, and control of equipment and hunting methods (i.e., Bartlett, 1949; Hunter, 1957; Longhurst, 1957; Murphy, 1969; and Jenkins, 1970). Most states have followed deer management programs based on mature males and antler size, with the use of various numbers of "antlerless" permits allowing the take of any deer to augment the kill to approximately the desired level. This is the program followed in the state of Michigan (Bennet et al., 1966). Many variables influence the size of the kill in such a system: the number of hunters and the weather conditions for hunting are just two of many. Whether the season opens on the weekend versus a weekday has a major impact on the number of hunters in the field (Hunter, 1957) and the size of the kill (Murphy, 1966). Michigan has opened the season on November 15 in recent years, without regard to day of week. Efforts to obtain a kill-per-unit effort, comparable to the catch-per-unit effort of commercial fisheries (see Ricker, 1975; Seber, 1973), have not been very successful (Eberhardt, 1960).

The influence of a variable kill due to the usual practice of fixing seasons on a model such as this is of considerable interest. What is the loss in yield that must be accepted in order to prevent overexploitation and possible extinction? Random normal variations can be used to adjust the kill, but the magnitude of the variation typical for statewide deer programs is difficult to determine. To obtain an estimate, the statistics for deer kill in Michigan were examined (Bennett et al., 1966; Ryel, 1974; and Hawn, 1974). Three regions are included in the data: Region I, the upper peninsula; Region II, the upper part of the lower peninsula; and Region III, the southern part of the lower peninsula. Data from each region were analyzed separately, and only data from the regular gun season were included. Regulations, with the exception of the number of antlerless permits, have been reasonably similar since 1958, and data were included up through 1974 for a seventeen-year sample. Since all hunters have a deer license for bucks, the total kill of bucks was regressed on the total number of hunters. However, the antlerless deer kill was regressed on only the antlerless permits. It was assumed that there would be positive correlations between the number of hunters or antlerless permits issued and the respective kill of bucks and other deer. The assumption proved true, because positive slopes were obtained. It was further assumed that the variance was proportional, rather than equal. This assumption was necessary in order to relate the rather different magnitudes of kill in the three regions and the low numbers killed on the George Reserve. If the standard deviation is expressed as a percentage of

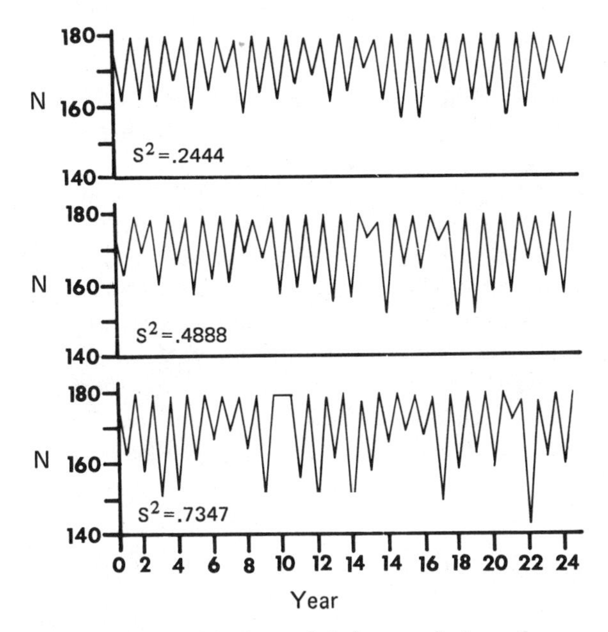

FIG. 9.3. Simulations with deterministic population change model and stochastic kill: top, with approximated variance in kill in the state of Michigan; middle, with variance doubled; and bottom, with variance tripled.

the mean, a range of 11.74 to 33.51 percent is obtained from the six regressions calculated. The mean percentage was 24.44, and it was accepted as reasonably typical of a statewide deer program such as that of the state of Michigan.

A series of simulations was run with the deterministic model in which the kill was adjusted by multiplying it by a random normal number with a mean of one and a standard deviation of 0.2444. Other parameters used in the previous series of simulations were retained. Average kill was set at 10 percent of the population (equilibrium kill is 16.15 with a posthunt population of 163.15), a crude estimate of the average kill in most white-tailed deer states. A remarkable stability emerged, despite the variation in the kill (fig. 9.3). Although the prehunt population varied with the size of the kill, the compensatory response of the deterministic population model resulted in a return to essentially the same peak population. Thus, the population was able to stabilize the peak population at equilibrium with kills ranging from 8.88 to 23.42. Outcomes were similar when the variance was doubled and tripled. The range in kill (and hence, size of the posthunt population) varied substantially, yet the peak population was readily returned to a nearly stable level.

This kind of simulation could be extended to various levels of exploitation, but such an exercise seems pointless. The three examples are sufficient to demonstrate that the introduction of a random variation in the kill has relatively little effect upon population fluctuation if the population change is deterministic and compensatory in the manner of the mean values derived in this study. Obviously, random variation in the kill would lead to a greater probability of extinction as the average kill approached MSY, since abnormally high kills would reduce the population below MSY. But the essential point of this set of simulations is that the compensatory aspect of the empirical, deterministic population change model of the George Reserve deer herd results in great homeostasis in the face of fluctuating kills.

Stochastic Population Change Model and Stochastic Kill

Simulations were run using stochastic population change and stochastic kill data similar to that of the state of Michigan (table 9.1). First extinctions occurred at a kill of forty-four per year, and the extinction rate rose rapidly with increased kill. Obviously, allowing both variables to fluctuate randomly greatly increases the sensitivity of the population to overexploitation; such fluctuation is most likely the case in nature. On the basis of these simulations, 87.8 percent of MSY could be taken with FRY without extinction occurring. Achieving MSY or FRY is strongly dependent upon successfully stabilizing the kill. Since stabilization, at least in principle, is an external variable amenable to management, it would aid agencies with such goals to give attention to stabilizing the kill.

Periodic Harvests and Yield

Walters and Bandy (1972) suggested that in big game populations a greater yield might be obtained from harvesting once every several years instead of annually. They further predicted that gain in yield from periodic harvest would be greatest for species with high rates of increase; e.g., deer would give greater yields than elk with periodic harvest. The authors reached this conclusion based upon the assumption that reproductive potential increases with age and that

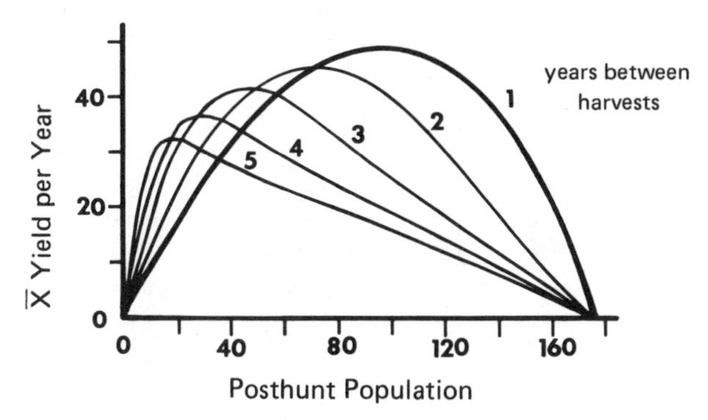

FIG. 9.4. Relationship of average annual yield to varying intervals between harvests

periodic harvests would allow a greater number of intermediate-aged animals to survive (on average) than would annual harvests.

The proposition can be tested with the George Reserve empirical model since recruitment integrates age structure, reproduction, survivorship of young, and population size—the important parameters of Walters and Bandy's model. The effects of periodic harvests can be determined by comparing the average yield of harvests at different intervals of years with the average yield obtained from annual harvests. The results (fig. 9.4) show that periodic harvest will reduce MSY and SY at posthunt populations of about seventy-eight and greater. Increasing the interval between harvests further depresses the average annual yield.

Therefore, it can be concluded that periodic harvests cannot increase MSY, nor can they increase average yield over SY if the population is at the peak or on the right descending segment of the yield parabola. If the regression of recruitment rate on posthunt population size is negative, periodic harvests will more greatly reduce the maximum possible yield than will annual harvests. Furthermore, contrary to Walters and Bandy's model, the reduction in yield will be greater for species with high reproductive rates because the higher the *a* value of the recruitment rate regression, the greater (i.e., more negative) the slope given comparable values of K carrying capacity. Time-lag effects (chap. 11) might lower the magnitude of the reduction of yield with periodic harvests, but it is inconceivable that time lags would reverse the trend. Furthermore, time lags would create the danger of vegetation destruction and a lowering of K carrying capacity, further reducing the long-term potential for yield.

There may be other management objectives, besides obtaining yield, that would favor periodic harvests: reduction of the disturbance caused by hunters on the populations, reduction of agency costs for special hunts, and achieving population control in areas such as parks that have high, nonconsumptive esthetic values are a few examples. Furthermore, with populations held to very low densities (i.e., on the left-ascending segment of the yield parabola in fig. 9.4), periodic harvests can increase the average yield over SY with annual harvests. As the posthunt population size following the harvest decreases, the average annual yield increases with increasing intervals between the harvests.

Yield in Relation to Sex Ratio

In theory, sex ratio adjustment should have a powerful effect on population performance. Skewing the population in favor of females in a polygynous species should increase rates of population growth and, hence, possible rates of exploitation. Skewing the population towards males should have the opposite effect. However, these expectations fail to account for the finite resources with which K-selected species live. The purpose of this section is to examine the ramifications of density dependent effects in the George Reserve deer population in terms of sex ratio manipulation and yield. Specifically, can MSY be increased by shifting sex ratio in favor of females?

Evidence presented in chapter 6 showed that sex ratio of the posthunt population had virtually no effect on recruitment. Recruitment rate was much more closely correlated with females only than with males only or both sexes combined. This result was obtained despite a relatively wide range of sex composition (43.9 to 70.5 percent female) over the nineteen years of reconstructed population data. Furthermore, an exhaustive examination of residuals failed to show any relationship of proportion of males in the population to recruitment rate. Thus, it was concluded that recruitment was determined by the number of females in the population, independent of the number of males over the range of posthunt populations observed. The conclusion would make sense if female-female competition for resources were much more direct than female-male competition; hence, the importance of the question of niche separation between the sexes (chap. 14).

Given that females alone are related to the recruitment and

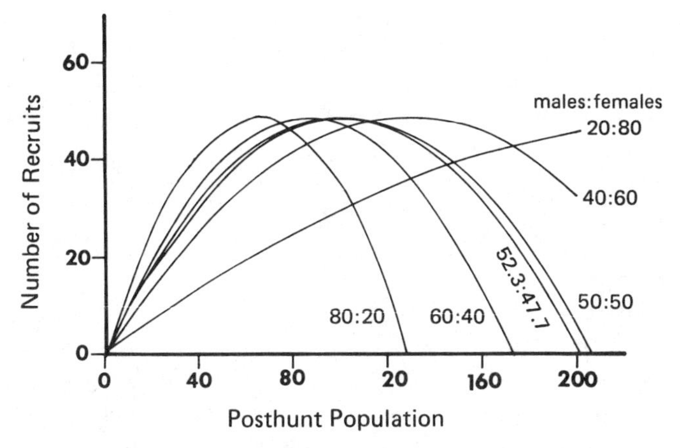

FIG. 9.5. Equilibrium recruitment curves at various sex ratios (females to males) as projected from recruitment based upon the females only regression of figure 6.1

that the relationship is negative, the impact of sex ratio variation can be determined by deriving the recruitment based on females only and adding variable numbers of males to achieve the desired sex ratio. These results would give mean equilibrium values, assuming the population size and sex ratio were controlled by hunting removals of a period of years. The results for the observed sex composition (52.26 percent female) and various, arbitrarily selected sex compositions are given in figure 9.5. Note that the figure includes recruitment only, excluding population change and factors like chronic mortality.

The first point of interest is that it is not possible to increase the MSY by adjustment of the sex ratio (note that all of the peaks are of equal height). A second interesting result is that sex ratio adjustment can achieve the same recruitment results at different, total posthunt population sizes. Thus, the MSY of the observed sex composition, which occurred at a posthunt population of ninety-nine, could be duplicated by a population of sixty-two if the sex composition were adjusted to 80 percent females. If, however, the population were simply reduced from ninety-nine to sixty-two without adjustment of the sex composition, a substantially smaller recruitment would be obtained (read up from 62 to the 52.26:47.47 sex composition curve of fig. 9.5). Stabilization at a given posthunt population size, of course, requires a removal of each sex equal to the recruitment (on average), so yield in numbers equals recruitment at the lower population sizes where chronic mortality is not present. It seems, therefore, that no increase in MSY can be obtained by adjusting sex ratio, and there is no advantage in discriminating between the sexes once the population has reached the peak of the recruitment parabola. However, the results suggest that MSY may be obtainable at a lower population size if the sex ratio is adjusted in favor of females. A corollary would be that the average age of the population would be lower, as would the age of animals in the kill.

At first it may seem as if the results are due to unsafe extrapolation of the recruitment rate on the number of posthunt females regression line. Note, however, that these results hold over the range of actual posthunt populations observed. Questionable extrapolations might occur at the higher posthunt populations, i.e., those on the right of figure 9.5. However, the expected values would be lower than those shown and at some point in increasing population size, competition between the sexes might become more pronounced.

It is of interest that the MSY of males cannot be achieved by shooting males only. This can be demonstrated by assuming that the deer population of the George Reserve is being managed for MSY. Then, instead of removing numbers of either sex equal to the number of recruits, males only are removed to equal the number of male recruits. As the population increases, chronic mortality will begin for females but not for males, which are still heavily harvested. The outcome of such a change in management is given in table 9.2, where recruitment is determined from the female only population regression. When managed for MSY, the equilibrium yields were 48.69 total and 26.00 males. Switching to a males only kill leads to an equilibrium six years later with an equilibrium yield of only 2.94 males. This is substantially fewer males than were taken at MSY. Since the number of males in the posthunt population remained unchanged (47.25), the sex composition of the posthunt would have shifted from 47.74 percent male at MSY to 31.85 at the new equilibrium. It should be apparent that this result occurs because the female segment of the population grows toward K carrying capacity and recruitment declines accordingly, even though males are being removed.

If the linear extrapolation of the females only regression equation is questioned, the total population regression can be used, but the outcome is similar (table 9.3). Equilibrium was reached in eight years, and the new equilibrium yield was 6.79 males. One could further elaborate the model by incorporating other changes like the

TABLE 9.2. Population Projections Assuming Removal of Numbers of Males Equal to Male Recruits as Projected from the Posthunt Female Regression. Projections Start (Year 0) at MSY.

Year	Prehunt Population	Total Kill	♂♂ Kill	♀♀ Chronic Mortality	Total Posthunt Population	Posthunt ♂♂	Posthunt ♀♀	Total Recruits	♂♂ Recruits
0	147.69	48.69	26.00	0.00	99	47.25	51.75	48.69	26.00
1	147.69	26.00	26.00	0.00	121.69	47.25	74.44	39.76	21.33
2	161.45	21.33	21.33	0.00	140.12	47.25	92.87	18.42	9.88
3	158.54	9.88	9.88	1.58	147.08	47.25	99.83	7.69	4.13
4	154.77	4.13	4.13	2.37	148.27	47.25	101.25	5.62	3.02
5	153.89	3.02	3.02	2.53	148.34	47.25	101.09	5.50	2.95
6	153.84	2.95	2.95	2.54	148.35	47.25	101.10	5.48	2.94
7	153.83	2.94	2.94	2.54	148.35	47.25	101.10	5.48	2.94

TABLE 9.3. Population Projections Assuming Removal of Numbers of Males Equal to Male Recruits as Projected from the Posthunt Total Population Regression. Projections Start (Year 0) at MSY.

Year	Prehunt Population	Total Kill	♂♂ Kill	♀♀ Chronic Mortality	Posthunt Population	Total Recruits	♂♂ Recruits
0	147.69	48.69	26.00	0.00	99.00	48.69	26.00
1	147.69	26.00	26.00	0.00	121.69	40.05	21.48
2	161.74	21.48	21.48	2.77	137.49	38.28	20.53
3	175.77	20.53	20.53	1.33	153.91	26.20	14.05
4	180.11	14.05	14.05	3.35	166.06	13.77	7.39
5	179.83	7.39	7.39	5.67	166.77	12.93	6.94
6	179.70	6.94	6.94	5.84	166.92	12.76	6.84
7	179.68	6.84	6.84	5.86	166.98	12.70	6.81
8	179.68	6.81	6.81	5.87	167.00	12.67	6.79
9	179.67	6.79	6.79	5.88	167.00	12.67	6.79

sex ratio of offspring, but the basic pattern will remain unchanged and the conclusion that MSY of males cannot be obtained by killing only males will stand. Note that at the new equilibrium, the males remaining in the posthunt population do not figure into the result beyond the simple number necessary to breed the estrous females. Thus, the number of males in the posthunt population could be further decreased by a greater kill of males in one or more years (i.e., above the number of male recruits), thereby further decreasing the ratio of males to females in the posthunt population without influencing the equilibrium yield of males. These results are consistent with the experiences of most states employing continued, bucks-only management, producing low kills of males (mainly young) and low ratios of males to females in the living population.

Considerable stock has been placed on the importance of bucks-only hunting in the recovery of deer populations during the restoration period following the turn of the century. It is clear that shooting only males would not stop population growth as long as there were sufficient males for breeding. The female segment of the population would continue to grow to equilibrium near K carrying capacity as in the examples given in tables 9.2 and 9.3. Thus, the assumption that bucks-only laws encouraged recovery of deer populations is probably correct. But once MSY posthunt population is reached, the advantage disappears. Females are stabilized by chronic mortality, assuming hunters are actually not shooting females by accident or design.

The conclusion of this analysis is that MSY cannot be increased by manipulation of sex ratio; skewing the sex ratio *in either direction* leads to a reduction of yield, unless it is also accompanied by a compensating shift in posthunt population size. However, MSY can be achieved at lower posthunt populations by adjustment of the sex ratio. In terms of yield of numbers, there is no value in trying to adjust sex ratio in the harvest once MSY posthunt population is reached. Achieving MSY of one sex requires simultaneous MSY of the other sex. Perhaps it is best to regard these as tentative conclusions. The results were obtained from uncontrolled shifts in sex ratio. It is hoped that in the future, deliberate manipulations of sex ratio on the George Reserve will give final verification. Therefore, while caution seems in order, the results obtained are clear enough to cast considerable doubt on traditional thinking about sex ratio manipulation in deer management. Note, however, that in dealing with species for which correlations of recruitment rate with population size are better for both sexes combined than for females only,

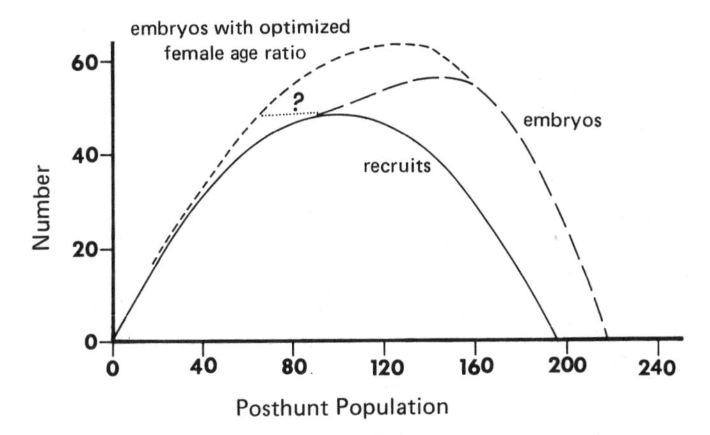

FIG. 9.6. Comparison of number of embryos (interrupted line) to number of recruits (solid line) and number of embryos obtained from optimizing age ratios of females (dashed line). The dotted line with a question mark suggests a possible recruitment curve with optimized age ratio of females.

the results might not hold. Thus, apparent niche separation by sex (chap. 14) becomes an important variable in the population dynamics of polygynous ungulates.

Yield in Relation to Age Ratio

As with sex ratio, one may question whether the sustained yield can be increased by manipulating the age structure of the population by differentially harvesting from the age-classes.

First, because yield is dependent on recruitment and the recruitment is produced by females, it can be concluded that adjustments of age structure of males will have no influence on yield as long as a sufficient number of sexually mature males are present for breeding. Although this number is not known, it cannot be many more than four to eight for the George Reserve, depending upon the size of the female population.

Second, it can be concluded that the adjustments of age ratio of females will not increase yield if the population size is to the right of the peak of the recruitment parabola (fig. 6.6). This is because the potential recruits (embryos *in utero*) already exceed the number of actual recruits limited by density effects (table 8.1, fig. 9.6). Increasing the number of embryos will only increase the amount of prerecruitment mortality. Yield, which is dependent upon recruitment, will remain unchanged. To the left of the parabola, nearly all embryos survive and recruitment is dependent on the number of embryos. If the number of embryos can be increased by changing the age ratio, then it might be possible to achieve higher yields. Therefore, the concern lies with females at population sizes below MSY.

In theory, the number of offspring from a given population can be maximized by selecting for removal those females with low reproductive values (Fisher, 1958)—the very young and very old. In practice, fawns can readily be identified in the field and selectively shot. Yearlings are large, particularly under the favorable growth conditions at low population size (i.e., to the left of MSY on the recruitment number parabola), and even experienced workers would have difficulty distinguishing them from older females under field conditions. It is doubtful that yearling females could be effectively selected; certainly the average hunter could not do so. Thus, although yearlings have a somewhat lower pregnancy rate than older females, they have to be included with that group from a practical standpoint.

Although pregnancy rates in older females remain high until old age, their reproductive value drops because of their greater probability of dying. Ideally, one would allow females to remain in the population until about ten to eleven years of age. Unfortunately, distinguishing age accurately in the field is virtually impossible. A few decrepit individuals could be distinguished, as could some of the smallest yearlings, but by and large all Y+Ad females would have to be treated as a group.

Because the number of embryos is equal to the number of recruits on the left of the recruitment parabola, the objective is to maximize the number of embryos produced by an equilibrium population. The strategy, therefore, would be to increase the proportion of the yearling and older segment of the female posthunt population, because of its higher embryo rates. The increase could be achieved by differentially killing female fawns. However, enough female fawns must be left to replenish the older age-classes and enough older females must be killed to prevent chronic mortality. Since the age of older females cannot be distinguished in the field, removals of older females must be great enough to ensure that, on average, all are killed before they reach the age of maximum longevity or 13 years. The mean life expectancy would be six years if one assumes a linear survivorship, a reasonable assumption since vulnerabilities of Y+Ad are reasonably similar (fig. 5.2).

A linear regression that crosses the x-axis at twelve years (not the longevity of thirteen years because the starting point is age-class one rather than zero) and has an area under the regression line equal to number in the Y+Ad population can be used to determine the number of female fawns that must be spared. This process is similar to that used in chapter 7 to estimate population size at K carrying capacity.

For example, at a posthunt population of ninety, there are 47.04 females at equilibrium—17.56 fawns and 29.48 older females. Thus, 37.33 percent of the posthunt female population would be fawns when shot by the selectivities shown in fig. 5.2. This female population would produce 22.47 female embryos and 22.51 female recruits (the difference is due to random errors in equation fits and is not considered to be real). Equilibrium of females at a total posthunt population of ninety would require killing 4.95 fawn females and 17.56 older females (total kill of 22.51 equals 22.51 recruitment). If a greater proportion of fawns were killed, the age ratio could be shifted higher, and a greater number of embryos would be obtained. Arbitrarily, the number of fawns remaining could be decreased until the number would be balanced by the linear survivorship of the older females. Assume that 10 fawn females were to be left. This would leave 37.04 older females in the posthunt equilibrium population. However, 37.04 divided by six, the average life expectancy of a linear survivorship, gives 6.17—the number of fawn females consistent with equilibrium of 37.04 older females. Leaving 10 fawn females would be more than necessary, since the actual number that would balance is 6.72 (i.e., 47.04 minus 6.72 equals 40.32; 40.32 divided by 6 equals 6.72). The posthunt population containing 6.72 fawns (14.29 percent), and 40.32 older females would produce 3.92 and 23.30 female embryos respectively for a total of 27.22 female embryos. Assuming that all survived to be recruits, 20.50 female fawns and 6.72 older females would have to be shot to bring the posthunt number of females back to the equilibrium number, 47.04.

Adding male embryos to the female embryos gives a total embryo count of 58.42, a number that could be produced at equilibrium with a posthunt population of ninety by maximizing embryos through adjustment of age ratio of females. The count obtained from the age ratio actually occurring on the George Reserve at an equilibrium posthunt population of ninety is 48.19. With selective harvesting, a gain of about 10 embryos could be obtained by optimizing the age ratio of females in the posthunt population size.

Similar calculations for other posthunt population sizes give the maximum embryo curve shown in figure 9.6. It can be seen that adjustment of the female age ratio can give a substantial increase in the number of embryos produced, but the survivorship of the additional embryos cannot be determined. From the observed embryos and survivorship, prerecruitment mortality began at a posthunt population of about ninety-five. The shape of the realized recruitment (dotted line in fig. 9.6) cannot, however, be determined. It seems likely that it would increase greatly over the peak value of the parabola, for such gains, if any, would not be very large. However, it seems likely that MSY could be obtained at a lower population size by adjusting female age ratios.

The age ratio results parallel those obtained for sex ratio. On balance, the gains from being selective of sex and age are minor. It would require an unusual situation, where control of harvest and the data base on population performance were exceptionally good, for the gains to be noticeable. For broad-scaled management, there seems to be little point in attempting to adjust either the sex or age ratio. Posthunt population size is the major variable, and MSY can be obtained by shooting deer as encountered at random.

CHAPTER 10

Carrying Capacity and K-Selection

Among theoretical ecologists, there has been a unanimity of opinion that carrying capacity means K of the logistic equation. However, the term has come to mean many different things in wildlife management (Edwards and Fowle, 1955).

Dasmann (1964*a*) has pointed out four different usages: (1) the maximum number of animals an environment will support, the K of the logistic growth curve, which he terms "subsistence density"; (2) the population level at which MSY is obtained, commonly used in range and wildlife management, which he terms "optimum density"; (3) the carrying capacity used by Errington (1934) to indicate the population level that was relatively immune to predation, which he terms "security density"; and (4) the maximum number of animals that can be crowded into a given space due to behavioral characteristics (such as the number of territorial pairs of breeding birds), which he terms "tolerance density." The terms proposed by Dasmann clarify the situation considerably, since each is somewhat different and for precise communication these differences must be distinguished. In fact, the term carrying capacity has become so ambiguous that it is useful only in the most general sense.

Dasmann's terms can be related to the population model of the George Reserve deer herd (fig. 10.1). Subsistence density occurs at K of the population change curve, while optimum density is under the peak of the population change parabola and is equivalent to MSY. Neither security nor tolerance density is operating on the George Reserve deer population, but they can be related in theory. Security density is that point on the population growth curve at which animals are forced out of a secure habitat and become susceptible to predators. Below this level, predation is minor. Errington has presented the relationship as shown in figure 10.2, where

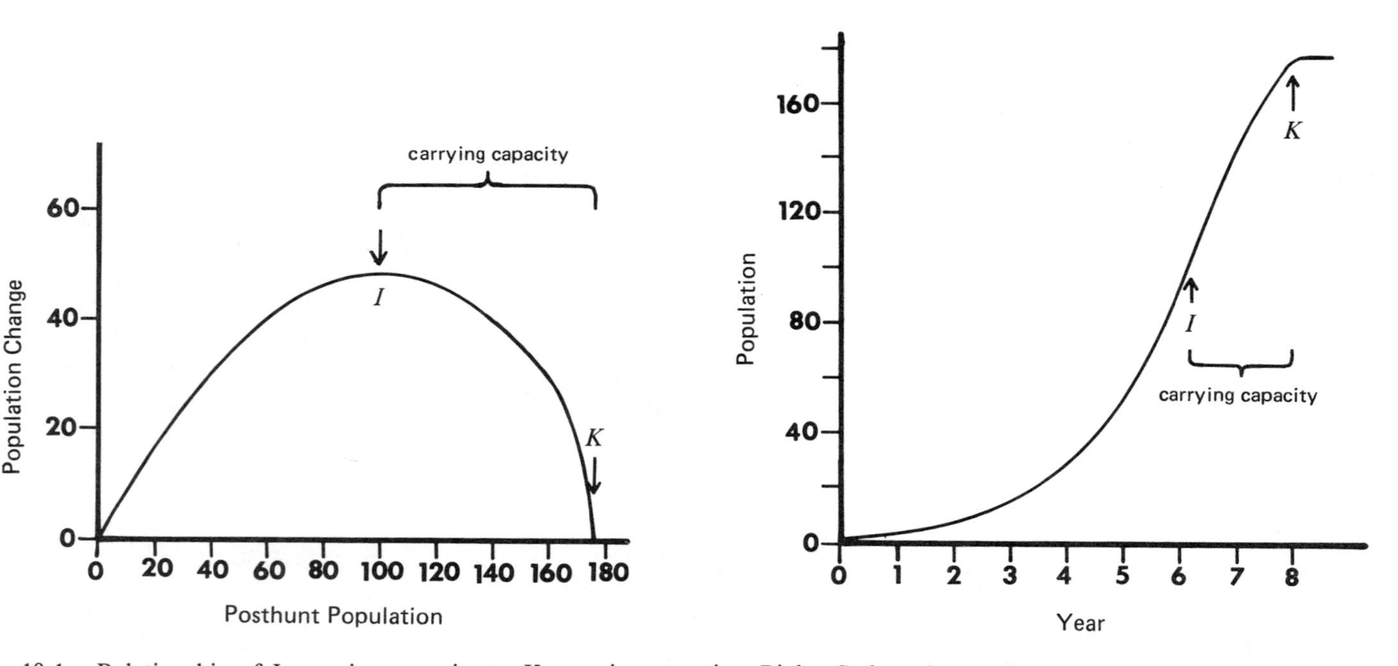

FIG. 10.1. Relationship of I carrying capacity to K carrying capacity. Right, S-shaped growth curve generated by the George Reserve deer population model. Left, population change parabola.

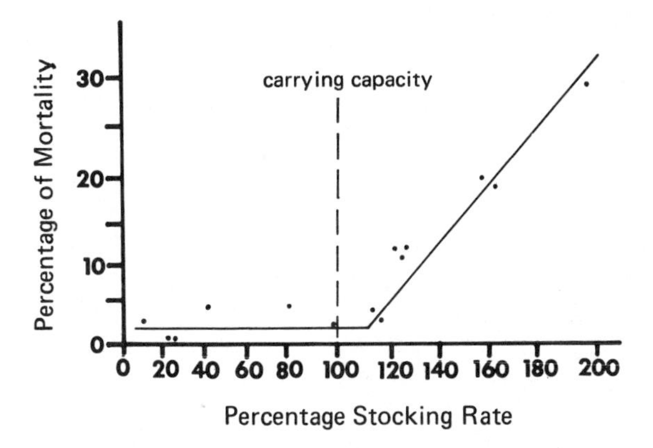

FIG. 10.2. Relationship of bobwhite quail density to rate of predation (after Errington, 1934)

predators capture few animals at low prey density, but capture an increasing proportion of them as the "threshold of security" is exceeded by greater prey numbers. Below this threshold, a low and fairly constant rate of predation would have little influence on population growth. Thus, if one started with a very low population in a good environment, geometric population growth should occur up to this threshold. Beyond this level, mortality would become progressively more intensive. Therefore, the threshold of security or carrying capacity of Errington occurs at the inflection point of the S-shaped growth curve. It is comparable to I in figure 10.1 (i.e., that population giving MSY) or the optimum density of Dasmann. Furthermore, under the Errington model of increasing predation, population stability would be achieved when the number of animals removed by predators was balanced by annual recruitment. This population level, unconsidered by Errington, would be located somewhere on the ascending arm of the curve in figure 10.2 and would be comparable to K carrying capacity in figure 10.1. The fundamental difference is in the point of reference, it being the inflection point in the Errington model. Thus, the carrying capacity of Errington is equivalent to MSY, while the zero population growth, unmentioned by Errington, would be equivalent to K. Also, the important factor in regulation is predators rather than food.

It seems that carrying capacity based upon behavioral characteristics of a species (tolerance density) also should be at MSY. This conclusion is based upon the following reasoning, which is

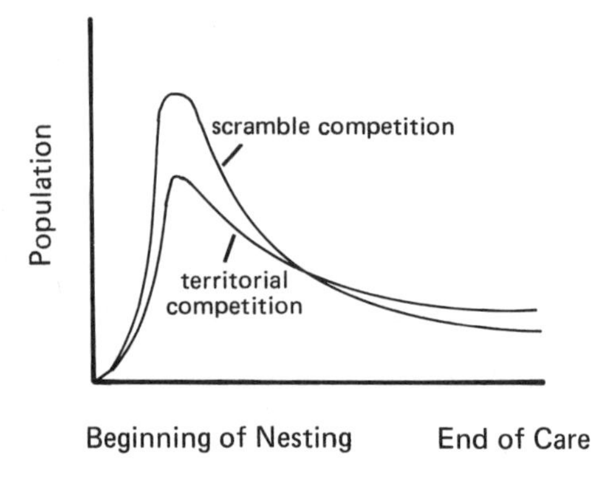

FIG. 10.3. Hypothesized relationship of outcome of scramble competition versus competition for territories in pair territorial birds

presented in terms of the well-known pair territorial behavior in birds (Brown, 1969) because resource territories in ungulates are uncommon and poorly studied. Most territorial behavior in ungulates involves male-male competition in a polygynous breeding system and, as such, has no importance in population regulation. All females are bred. Some small, forest-dwelling ungulates like brockets and duikers apparently have pair territories, but since little is known about them, birds provide a better example. It is assumed that territorial behavior maximizes the number of young fledged per pair. This must be true for selection to have favored territorial behavior (conventional competition) over open, or scramble, competition for resources, nesting sites, and the like. For territoriality to have evolved, pairs that defended territories must have been more successful in rearing offspring than their nonterritorial counterparts. Figure 10.3 shows the relative, evolutionary relationship of the territorial versus nonterritorial pairs. Because some birds are excluded from breeding, the eggs produced by breeding birds in a pair territorial situation would be expected to be fewer than if all birds were allowed to breed. However, adequate resources are protected by territorial birds, and only the more secure nesting sites are occupied. Competition for food and the exposure of nestlings to weather and predators must have produced a higher mortality rate for pairs in open competition than for territorial pairs occupying secure sites.

Given the advantage of the territorial breeding system in leaving a greater number of surviving offspring, selection would favor expansion of the size of territories of some pairs at the expense of others being excluded from the breeding effort. This process would continue until the cost-benefit ratio between additional young reared and the cost of maintaining a larger area with a greater perimeter was balanced (Brown, 1964). Ultimate limitation of territorial size may be set by a physiological constraint (i.e., the pair cannot process additional resources to produce more young) or by an energetic-nutrient balance where the cost of delivery of resources, defense of territory, etc., begins to come at the expense of production of young. This would seem to be an optimization process, since selection would favor the pairs that were most successful in optimizing the various costs and benefits to produce the greatest number of surviving young. The population consequences of this optimization would be equivalent to MSY. Variation in resource quality between seasons, population size, age and experience of pairs, and other factors would complicate the model but would not change the essential relationships.

Therefore, the population consequences of resource territories would be as depicted in figure 10.3. The size of the breeding population in the breeding habitat should expand to the MSY level but not beyond. Expansion beyond this level would result in lowering the production of young, and selection should favor exclusion of animals from the breeding population. The assumption is based upon a homogeneous breeding habitat with discrete boundaries; if this were not the case, pairs with lower breeding success on marginal territories would cause optimization of the total breeding habitat, although the success of pairs in the marginal areas would not equal the success of pairs in the best habitat. Thus, heterogeneous breeding habitat involves the definition of the unit of area in which the population is studied but not the question of optimization, of territorial size, or the resultant breeding population size.

The optimization of numbers of surviving offspring seems to be the usual case for breeding populations in breeding habitat, and carrying capacity set by behavioral mechanisms would be comparable to the MSY point shown in figure 10.1. Treating the breeding population separately from the total population is a specifically defined case including only the period of reproduction. Over a longer period of time, and including nonbreeding adults and offspring of breeding adults, some sort of regulation at K carrying capacity must occur. Inclusion of all animals over a complete generation will give a carrying capacity equivalent to K, while inclusion of only the breeding population and territories should approximate the carrying capacity at MSY.

The various kinds of carrying capacity that have been defined can be resolved into two sets relative to an S-shaped population growth curve: that set associated with the zero growth point (K), and that set associated with the inflection point (I) that gives MSY. The confusion of terminology, therefore, stems from (1) failing to distinguish these two fundamentally different points on the S-shaped population growth curve, (2) considering subsets of the population or generation time, and (3) using labels that emphasize causality rather than the point on the S-shaped population growth curve. To be logically consistent, the only definition of carrying capacity that should be used is that of the maximum number of animals an environment will support or K. K is the endpoint of a functional response of a population where reproduction and mortality balance out at maximum population size. The maximization of yield (MSY) does not change the maximum number of animals that the environment will support—it merely holds the population below that endpoint. For example, the inflection point in the Errington model (fig. 10.2) is not the maximum number the population will achieve. That maximum will occur at some higher population level, where mortality due to predation is balanced by annual recruitment of young. The population is still below K, and K would not be reached unless predators were reduced. The fact that behavior may limit the number of breeding birds does not alter the fact that the breeding birds, plus the young recruited, plus the nonbreeding birds comprise a population that will come into balance with the capacity of the environment to support that population. It should be clear, therefore, that these other definitions of carrying capacity involve a set of conditions below K in the functional relationship of population to environment of special biological interest. While the relationships are important and biologically fascinating, they do not represent the capacity of the environment to support the maximum number of animals and should not have been labeled "carrying capacity;" the fact that they have has confused both the term and the concept involved. At this time, it is unlikely that the single definition of carrying capacity being K can be effectively established in the wildlife field. Purity of the definition is much less important than making the discrimination between the important differences in the concepts.

There would seem to be no great difficulty in defining characteristics of the population relative to points on an S-shaped growth curve (fig. 10.1), and any population that has a growth form approximating an S-shaped curve could be related to this graphic model. To clarify communication, terms should be defined in reference to the graph, rather than to the environmental factor or factors responsible for regulation. The following definitions, which are illustrated in figure 10.1, are proposed: *Carrying capacity* is a general term for the capacity of an environment to support organisms, including all the area of an S-shaped growth curve lying between the inflection point (I) and the upper asymptote (K). *K carrying capacity* is the maximum number of organisms an environment can support, as indicated by the upper asymptote (zero growth point) of the S-shaped growth curve. *I carrying capacity* is the population size at the inflection point of the S-shaped growth curve which results in MSY.

The reason for designating I as the beginning point of carrying capacity is that it is at this point that the pronounced effect of environment on the population begins. In the George Reserve deer population, mortality of embryos begins here and increases rapidly as population exceeds I. This is not to say that all environmental effects are shown by the population above I—the declining embryo rates with increasing population size (fig. 4.8) occur below I—but it is at I where survivorship becomes important, and where the relative magnitude of environmental effects begins to increase rapidly with further increases in population size.

If desired, the causative factors that set the level of K or hold the population at I can be indicated by subscripts; for example, f for food, or p for predation. In some cases, one environmental factor is of overwhelming importance in population regulation, and a subscripting system would work well. However, in many cases the causative factors are numerous and interacting in effect. Furthermore, these factors are dynamic and frequently vary from season to season and year to year. Since they tend to be interrelated in compensatory fashion, increases in one factor frequently result in decreases or elimination of others. In the George Reserve deer herd, virtually all natural factors have been eliminated by human hunting. Yet it is certain that curtailment of the heavy hunting kill of recent years would result in an increase of natural factors in population regulation. In these cases, a subscripting system will be somewhat more ambigious but perhaps less confusing than the present usage of terms.

Do Ungulate Populations Track K Carrying Capacity?

In nature, the relationship of population to carrying capacity is obscured by variation of an unidentified origin. It is rare for an ungulate population to be consistently stable in the absence of artificial regulation by man. There are four posible explanations for the source of this variation. (1) The population shows pronounced fluctuations in the absence of variation in the environment. The implication is that this is an inherent outcome of the population parameters of the species and stabilizing the environment would not result in a stable population. (2) The population tracks K carrying capacity, but carrying capacity varies greatly over time and is independent of the influences of the population. In this case, the population is constantly moving in the direction of K carrying capacity but never achieves a stable equilibrium because K carrying capacity is constantly shifting. (3) The population overshoots a previously stable K carrying capacity and induces changes in K carrying capacity through the impact of the population itself. In the presence of the resulting time lags, even though the population may generally track K carrying capacity, a stable equilibrium is not achieved. (4) A combination of two or all three of the above cases. Because factors are operating simultaneously in nature, it is usually impossible to determine which of the above situations is occurring.

In previous chapters, the variables have been kept separate. Basic carrying capacity of the George Reserve has been quite stable, and time-lag effects have been treated separately. Therefore, population responses can be examined independently of changes in K carrying capacity, most of which are due to time-lag effects on the George Reserve. The latter will be covered in the following chapter. The question here is whether white-tailed deer or other ungulates would approach a stable equilibrium population at K carrying capacity if K carrying capacity were stable and no time lags were present. The question can be examined relative to the equilibrium tendencies discussed in chapter 8, as measured by the value of a of the recruitment rate regression or the L value of Maynard-Smith (1968). These values are measures of the tendency of the population to achieve a stable equilibrium at K carrying capacity, to overshoot K and achieve an equilibrium after dampened oscillations, or to fail to achieve equilibrium at all. Given the presence of a stable K carrying capacity and no time lags, L values greater than two indicate populations that will fluctuate independently of K carrying capacity. In essence, because of inherent population responses, these populations will not track (i.e., change in population size so as to move in the direction of K) K carrying capacity except in a very general way. L values of one to two show overshoot and dampening oscillations until a stable equilibrium at K carrying capacity is achieved. Such populations tend to track K carrying capacity but have inherent population responses that result in overshoot. Thus, if the population has the capacity to cause damage to the vegetation, time-lag effects are likely to result. As the L value approaches two, the tendency of the population to track K carrying capacity decreases and the tendency to overshoot increases. Conversely, as the L value approaches one, the tendency to track K carrying capacity increases and the tendency to overshoot K decreases.

The L value of the George Reserve deer population was 1.06. Thus, the inherent population responses have a slight tendency to overshoot and introduce time-lag effects, even though the population also has a strong tendency to track K carrying capacity. An L value of one or less leads to stability at K carrying capacity without overshoot. Such populations have a very high tendency to track K carrying capacity, and the tendency increases as L approaches zero. Most climax ungulates have a single offspring per year and would fall into this range of L values. For example, the tule elk population of Owens Valley, California has an L value of 0.3731 (McCullough, 1978). Again, low L values do not imply that the population will actually achieve equilibrium at K carrying capacity; only that the tendency is present, and in the absence of time lags and in a stable, highly predictable environment relative stability should emerge. In nature these conditions are seldom met. Even at equilibrium seasonal reproducers overshoot K on a seasonal basis. Populations with low L values in unstable environments should, however, tend to track the shifting K carrying capacity more closely than would populations with high L values. The L value, or the value of the regression of recruitment rate on population size, is but one of the ways this tendency can be expressed; remember that these are only approximations.

The relatively high rate of increase in white-tailed deer is probably related to their adaptation to subclimax vegetation, a patchy and unpredictable resource. The uncertainty of its location, amount, time of creation, and duration of usefulness means that such areas can only be exploited effectively if the species has relatively rapid population responses. These results can be expressed as the maximum number of young produced per year per female if the sex ratio is balanced. This is equivalent to clutch or litter size if one assumes a single reproductive effort per year. Approximately 2.52 is the maximum number of young produced per year which will achieve equilibrium. The value will vary somewhat by species because the degree of curvilinearity of the population change on population size regression will differ due to longevity.

Definition of K-Selected

Since MacArthur and Wilson (1967) first coined the terms, *r*- and K-selection have been useful in distinguishing between organisms with radically different biology—*r*-selected being those that maximize the rate of population increase and K-selected being those that maximize competitive ability in a crowded environment. Although MacArthur and Wilson discussed the terms in reference to the forces of selection on a given species over time in a changing environment, subsequent workers, for example Pianka (1970, 1972) and Stubbs (1977), have used the terms to classify species at essentially a given point in time, or the present. On a gross basis there tends to be a strong differentiation between species of one category or the other, but there are difficulties with this kind of classification, as have been pointed out by Wilbur et al., (1974) and Stubbs. All life history parameters are the result of selection, and they cannot be neatly sorted into *r*- or K-selection for a given species. For example, some species such as forest trees or sessile intertidal organisms are strongly K-selected in the adult stage but produce massive numbers of propagules and thus are *r*-selected reproducers. If the *r* to K spectrum is considered to be a continuum, it cannot deal with the dichotomies in different stages of the life cycle.

At least four (and perhaps more) categories are needed to classify the *r*-K relationship.

r-*strategists*. Those species which maximize rate of increase in all stages of the life cycle. They have short life expectancies and short generation time; they produce massive numbers of offspring, and give little or no parental care. They are semelparous (reproduce once in their lifetime) and show population responses that are largely density independent. Survivorship tends to be extremely low throughout the life cycle.

r-*K dichotomists*. Those species which have an *r*-strategy of reproduction but a K-strategy at some other stage in the life cycle, usually the subadult and/or adult, reproductive stage. They occupy two positions on the *r*-K spectrum, with the first stage being near the *r* extreme and the second being near the K extreme. They are iteroparous reproducers (reproduce many times in their lifetime). Offspring (or propagules) have largely density independent population responses while adults are highly density dependent. Survivorship is usually extremely low for offspring but very high for established adults.

r-*K balancers*. Those species which achieve a trade-off of rate of increase relative to parental care, life expectancy, and repeated reproductive efforts. They occupy a single position on the *r*-K continuum, but various species are broadly distributed in the middle range between the *r* and K extremes. These populations show some density independent and some density dependent population responses, with the latter becoming more important as the position falls closer to the K end of the continuum, and vice versa.

K-strategists. Those species which maximize survivorship and competitive ability at all stages of the life cycle, and nearly all stages of the life cycle are density dependent.

This classification is envisioned as having true *r*-selected and true K-selected species occupying a relatively narrow spectrum at each end of the *r*-K continuum, with the *r*-K balancers occupying the broad, middle range between them. In terms of numbers of species, *r*-selected will contain the overwhelming majority, with numbers dropping precipitously at the start of the *r*-K balancers and very gradually thereafter as one moves from *r* to K on the continuum. The distinction between the *r*-K balancers and the true K-selected species is of major interest, since it appears that the line between the two can be objectively defined, based upon the maximum value of *a* of the recruitment rate regression or L value that will achieve equilibrium (chap. 8). If one assumes a stable habitat and no time lags, populations which have an *a* of the recruitment rate regression which is approximately 1.26 or less will achieve equilibrium. Since K-strategists are those species that have life history parameters selected by K carrying capacity situations, the potential to achieve equilibrium seems a reasonable definition of a K-strategist species. Hence, K-strategists can be defined as those species with recruitment rate regressions in which *a* is equal to or less than approximately 1.26. Species with values of *a* greater than approximately 1.26 will not achieve equilibrium on the basis of their inherent population parameters. The implication is that these species are not totally K-selected; they show a population rate of increase sufficiently great to produce instability in population size.

They would naturally fall into the *r*-K balancer category. Species such as rabbits, rodents, and most birds are examples, even though they lie on the K end of the *r*-K continuum. They are density dependent to a considerable extent but still they have a high enough rate of increase to be inherently capable of overshoot in the absence of time-lag effects. Thus, the balance between rate of increase and competitiveness in a crowded environment is shifted towards rate of increase as compared to complete K-strategists. Population regulation in these species is much more dependent upon environmental forces. The existence of refugia and environmental predictability are much more important to *r*-K balancers in achieving stability than to complete K-strategists.

While in birds and mammals being large is virtually synonymous with being K-selected, being small does not necessarily mean a species is not K-selected. Bats and gulls, for example, appear to be K-selected despite their small size. However, in other respects their life history parameters are more as expected, such as a slow maturity and long life span. Most ungulates are K-selected. Exceptions might be the pigs (Suidae) which have large litter sizes. Among North American ungulates, the white-tailed deer is probably at the *r* end of the true K-selected species. It is virtually a rule that under good conditions, subclimax North American ungulates (deer, pronghorn antelope, moose) have two offspring per adult female per reproductive effort, while climax ungulates (bison, bighorn sheep, mountain goat, muskox, caribou) and intermediate (elk) have only one. Like the true pigs, the piglike peccary (Tayassuidea) seems to be the exception: it is a climax animal, but has two offspring per litter.

Adaptation in Extremely K-Selected Species

Extreme K-selected animals are confronted with a problem in reducing the rate of increase since litter size cannot be reduced below one. Further reduction of rate of increase is dependent upon delayed sexual maturity (thereby extending the period between reproductive efforts) and loss of reproductive capacity before the end of the life span. Detailed reviews of reproductive patterns in mammals are given by Asdell (1964) and Sadlier (1969).

Delayed sexual maturity is a common adaptation in K-selected species, and in extreme examples (e.g., elephants, humans) it is associated with slowed growth rates and long juvenile periods of approximately fourteen years. In an animal that typically has a one-year delay in sexual maturity, the delay amounts to about 6 to 8 percent of the life span, while for humans it represents about 20 percent.

While the length of gestation period is an obviously important parameter in rate of increase, for the most part it is uncommon as a means of lowering the rate of increase in extremely K-selected species. One offspring per year is allowed by most gestation periods; the period seldom exceeds one year, even in the great whales. The exception seems to be the elephant, which has a gestation period of approximately two years. The elephant is probably the most highly K-selected living animal. Delays between production of offspring are common in K-selected species. Bears typically produce offspring every other year and, under poor conditions, at even longer intervals. Skipping reproduction in alternate years is a very common occurrence in ungulates under poor conditions.

An extended postreproductive life occurs in the most highly K-selected species. Williams (1957) and Alexander (1974) have argued that cessation of reproduction was selected because the advantages of investing further parental care in existing offspring outweighed those of producing further offspring and the risks involved. Ultimately this, or a similar argument, will account for the evolution of menopause. As pointed out by Cole (1954), Gadgil and Bossert (1970), and others, such selection will have population parameter consequences; in this case, the effect is that of reducing the rate of increase. Because population consequences are of such overwhelming importance in K-selected species that commonly live in intense, resource competitive situations, they have a pronounced effect upon the trade-offs of further parental investment versus additional reproductive efforts, parent-offspring conflicts, etc. Thus, while it is always imperative that *r*-selected organisms reproduce, in the lives of K-selected organisms it is often imperative not to reproduce, or even to discontinue reproductive efforts already begun.

A rough comparison of what these variables influencing reduction of rate of increase mean in terms of absolute numbers of offspring produced can be obtained by some simplistic assumptions. Under ideal conditions a female white-tailed deer has no delay in sexual maturity, produces a single offspring the first year and twins thereafter, and has a life span of thirteen years. A female could, therefore, produce twenty-five offspring in her lifetime. The bighorn sheep—with a year delay in sexual maturity, a single offspring, and a similar life expectancy—could produce twelve offspring, or about half of that of the white-tail. If we assume that the human female reaches sexual maturity at the age of fourteen, menopause at forty, and produces an offspring every other year, the total young produced would be 13, or very similar to the bighorn sheep, a K-selected ungulate which lives in climax situations. The white-tail, with one-fifth the life span of a human, would produce about twice the offspring. If humans had a biology similar to ungulates, reached sexual maturity at two years of age, and reproduced throughout the life span, the number of possible offspring would be nearly 70. Delayed sexual maturity would reduce the number by fourteen, intervals between reproductive efforts by thirteen, and menopause by thirty. The importance of a long post-reproductive life in lowering the rate of increase seems apparent and, on a comparative basis, seems to occur only in a few of the most highly K-selected species.

CHAPTER 11

Time Lags

We will now consider time lags in population responses. First, there is the lag time of responses within the constraints of the empirical population model given a change in management (e.g., increasing or decreasing the annual kill, fig. 9.1). These time lags are predictable from the empirical model. Given that the population is at equilibrium at a given level of annual removal, changing the level of removal by a given amount will lead to a new equilibrium point, and the length of time until the new equilibrium is achieved can be predicted. Furthermore, if stochastic processes are included in the model, the variation in time lag can be estimated empirically by numerous simulations. Finally, one can follow the error in tracking of population response to a set, but highly variable kill in a time sequence. Time lags of this kind present little problem because they are all predicted by the model.

Also encompassed by the model is the overshoot tendency expressed by the L value (chap. 10). Species with rapid rates of increase tend to overshoot, independent of time lags in vegetation response. Shifts in age structure accompany these time lags within the model (chap. 5). Achieving the equilibrium age structure may take longer than achieving the equilibrium population size. The second basic kind of time lag involves lags outside of the model, all of which are associated with population overshoot of K carrying capacity. It is valuable to examine the nature of overshoot because it is an important phenomenon in real world population ecology (Caughley, 1970). Even though management potentially could eliminate overshoot, it is often impossible to do so because of political, economic, or practical constraints, not the least of which is the lack of adequate information upon which to establish a program.

The overshoot of K by the George Reserve deer population occurred in the early 1930s and was quickly controlled by hunting, so that the impact became more subtle and spread over a period of

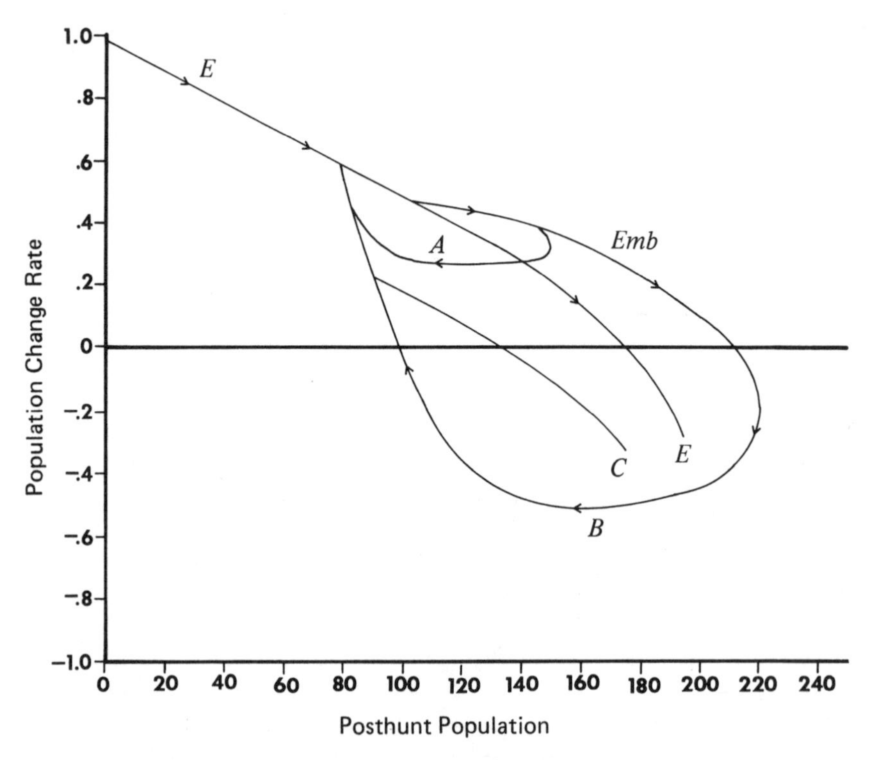

FIG. 11.1. Time lag model of rate of change in population under various conditions. Curve *E* represents equilibrium values. *Emb* represents overshoot due to embryos which fail to survive in equilibrium surviving to recruitment. Curve *A* represents the minor overshoot which occurred in the early years on the George Reserve. Curve *B* represents subjective estimate of the case of no population control by hunting. Curve *C* represents control until partial habitat recovery, then release from control.

years. A population "crash," typical of irruptive populations of large ungulates (for examples see Leopold, 1943; Leopold et al., 1947; and Caughley, 1970), did not occur on the George Reserve. The only data for that period were derived from the drive counts, which contain some error. Nevertheless, there are key relationships known for the George Reserve deer herd that serve as check points and allow an approximation of the actual case. The pertinent data for the George Reserve overshoot are contained in figures 3.1 and 6.2. The actual case can be modeled by simply smoothing the data points for the five-year moving mean for the drive counts in figure 6.2. This model is shown by curve *E*, *Emb*, and *A* in figure 11.1. If one starts with a low population (a virtual necessity for the overshoot phenomenon to occur), the initial recruitment rate will be the same as shown in figure 6.5. This is because virtually all young born survive below the MSY posthunt population of ninety-nine. Furthermore, recall that the early population growth on the reserve, the embryo rate data, and the recruitment data were consistent.

Above a breeding population of 100 is where the phenomenon of overshoot becomes apparent. If vegetation and deer population were in an equilibrium relationship, the recruitment rate would continue to follow the population change regression *E* in figure 11.1. Following the regression of these higher population levels is accomplished by increasing the mortality of young and beginning the chronic mortality of older animals, as discussed earlier. However, the actual recruitment in the five-year running mean of figure 6.2 follows the embryo rate of Y+Ad females (*Emb* of fig. 11.1). Thus, while at equilibrium state the embryo and other early losses would occur prior to recruitment age, in population overshoot these embryos were surviving to recruitment and were added to the previous population size. Because observed recruitment exceeded the expected (*Emb* curve versus *E* curve), the clear implication is that resources were in greater supply than they would be if the vegetation and the deer population were in equilibrium. The absence of this equilibrium is what produces the time lag, thereby allowing the overshoot of K.

If one started with a population of 99 animals in equilibrium with vegetation, and allowed them to build up, time lag would be unlikely to occur. In fact, this situation was demonstrated on the George Reserve. Once the population was reduced sufficiently to allow the vegetation to recover and come into equilibrium with deer feeding pressure, subsequent increase in the deer population followed the expected recruitment to population size relationship without an overshoot. Overshoot occurs only if the population has been held to a level substantially below MSY to allow "accumulation" of food resources (in a dynamic, living vegetation sense). A similar conclusion was reached in a survey of deer population overshoots in North America by Leopold et al. (1947). Introducing a population to a previously unoccupied range (Klein, 1968; Caughley, 1970) or suddenly creating a large amount of new habitat will produce the same result. The magnitude of the overshoot will be related to the magnitude of the accumulation of resources, and the depression of the resources and precipitousness of the population crash are, in turn, related to the magnitude of the overshoot. Thus, instead of a single curve of recruitment rate on breeding population for overshoot, there is a family of curves. Curve *A* of figure 11.1 is an approximation of the minor overshoot observed in the early years of the George Reserve. Even though shooting began, the posthunt population approached equilibrium carrying capacity (176), and the prehunt population exceeded zero recruitment (198 animals) twice (222 in 1934–35 and 214 in 1936–37). The latter peak was followed by another year of high population (fig. 3.1). These facts would account for vegetation suppression and subsequent declines in recruitment rate, despite reductions in the population size by hunting.

Recovery of recruitment rates (a reflection of the recovery of vegetation) began to occur when the breeding population fell below 90 animals and was achieved at 80 animals (fig. 6.2). Presumably the level at which recovery occurs varies with the magnitude of overshoot and the degree of the corresponding suppression of vegetation. The deviation of the overshoot of the curves in figure 11.1, therefore, is expressed primarily as an extension of survivorship of embryos to the right, rather than inflection upward. The recovery phases, by contrast, must be expressed primarily by deflection downward. Thus, the figure's curve, which approximates what happened on the reserve (*E, Emb,* and A) as the high population was gradually brought under control, represents a state of continuous adjustment during which equilibrium conditions were not reached. Recognize that this must be so, for if one stopped the control by hunting and the curve held, the population would build up to the mean recruitment-rate regression line and eventually stabilize at K—an impossibility if vegetation suppression has occurred. In fact, a population crash could occur only if the recruitment-rate curve dropped below zero. An obvious implication of these relationships is that accumulation of unused resources is relatively small in magnitude (the area lying above the mean regression line, *E* of fig. 11.1) as compared to the potential for suppression of vegetation by excessive deer numbers (the area lying below the mean regression line). Thus, minor overshoots have a disproportionately great impact upon the habitat. Or, to put it another way, damages produced by a small overshoot require a disproportionate reduction in population size for recovery to occur. Again, this result is consistent with the conclusion of Leopold et al.

A relatively small rate of hunting during the late buildup phase can introduce stability, but once the vegetation has been suppressed, the crash or reduction by hunting must be great enough to reestablish equilibrium conditions between herbivore and vegetation. If the crash reduces the herbivore population to a very low level, vegetation recovery can occur. If the herbivore population builds up again before complete recovery has occurred, a new equilibrium regression of recruitment rate would lead to population growth up to the reduced K. This would explain why many herbivore populations, following irruption and crash, build to a new K carrying capacity much lower than the original (Leopold et al., 1947). The pattern is particularly common with populations of exotics (Caughley, 1970). The reduction in carrying capacity can be modeled in figure 11.1 by assuming the new buildup curve is the same shape as curve *E,* but that the regression equation is lower by the same amount as the point of recovery on the family of curves represented by *A,* and *B* is below the mean equilibrium value. For example, curve *C* of the figure assumes that reduction of the posthunt population stopped at 90 posthunt animals and the population was allowed to increase to the reduced K. This would occur where curve *C* intersects the zero population change point, or a posthunt population of 133 in this case. The cost of the overshoot, in terms of loss of K carrying capacity, was a reduction of 43 animals (from 176 to 133). If, however, equilibrium conditions between vegetation and herbivore have been completely achieved prior to the second buildup, the mean regression line would be followed, as was the case with the George Reserve deer herd where recent minor buildups (fig. 3.1) followed the mean regression line.

If the herbivore population were held to a very low level for a longer period, so that vegetation not only recovered but accumulated, the stage would be set for a second overshoot. The very low, recent levels to which the reserve herd has been reduced are expected to have such a result. The current study plan of letting the herd build to 130, where it will be controlled, is expected to result in a slight overshoot, followed by the gradual adjustment of recruitment rate downward to equilibrium at the mean regression line. In fact, this is the prediction which is being tested by the next experiment. It is anticipated that the overshoot effect will be more slight than in the original buildup, because the short time at very low deer populations would not allow for the recovery of vegetation to its original condition. Although data are not available to verify the idea, it seems reasonable to suppose that highly palatable plant species that are susceptible to grazing pressure have been eliminated or reduced to a fugitive state on the reserve. Their rapid recovery is prevented by competition from more resistant plants established in their stead, as has been reported by Leopold et al. and Halls and Crawford (1960). Since palatable plants were more common prior to irruptions, it seems reasonable to suppose that they have the competitive advantage over unpalatable plants in the absence of deer feeding pressure. While some changes may be irreversible over practical, management time scales, the expectation would be that palatable species would increase in the absence of deer pressure. Because unpalatable plants replace palatable ones under deer feeding, the two categories must have similar ecological tolerances in other respects. The additional metabolic cost to the unpalatable species for developing protective structures or secondary compounds must be the critical difference that allows the palatable plants to win in competition with unpalatable plants in the absence of feeding, but the lack of such protection results in their losing in the face of heavy feeding. The complexity of ecosystem processes is illustrated by the paradox. The species of plants best adapted to survive are dependent for their success in competition with palatable plants on the presence of high populations of herbivores. Stable coexistence of palatable plants, unpalatable plants, and deer populations requires a reasonable balance between the three, a balance easily upset. Therefore, time lags result, and instead of a stable coexistence, a dynamic, oscillatory coexistence is observed.

While it is not possible to say precisely what the response of the George Reserve deer herd would have been in the early population buildup if hunting had not been instituted, the general relationship can be given as the most extreme of the family of curves (*B* in fig. 11.1). A hypothetical case of the history of George Reserve deer, assuming no control measures had been taken as derived from this model, is shown in figure 11.2. Here the population was allowed to grow unimpeded, to crash, and to secondarily recover to a new, lowered K; the deer population was never low enough to allow complete recovery of the habitat. An infinite set of histories for this herd could be generated by varying the control at various stages of the time-lag phenomenon. These are easily further elaborated by stochastic processes on the rate of population change and size of kill. However, the results of the deterministic model with no human controls are sufficient to show the basic responses to the time-lag model.

The studies were done in the absence of significant habitat manipulations. Indeed, it was the relative constancy of this variable which allowed the conduct of the experiment. However, the

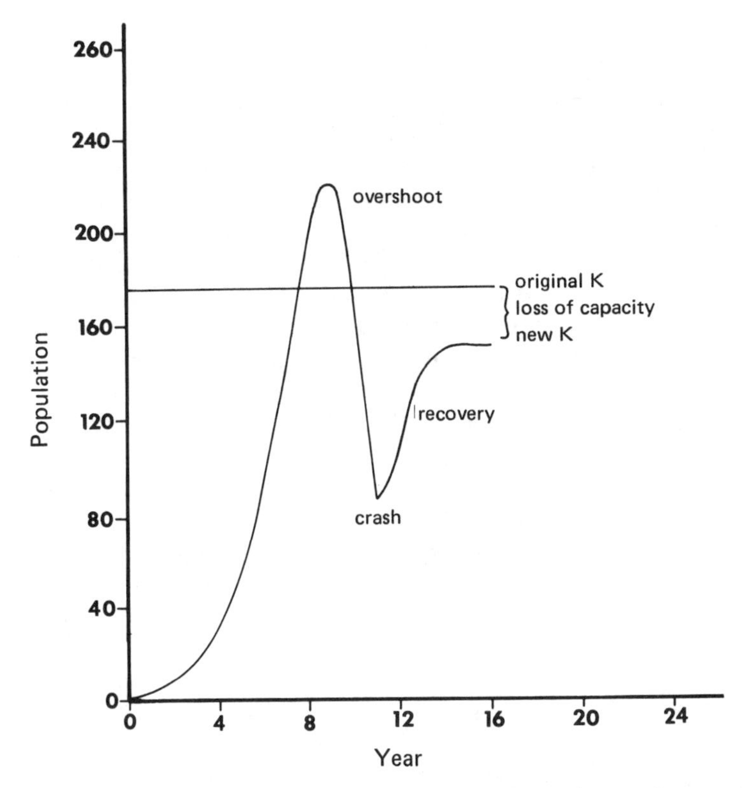

FIG. 11.2. Population history assuming no hunting and a starting population of one pair of adult deer

white-tailed deer is an inhabitant of subclimax vegetation, and in most (particularly forested) situations, habitat variation is a major comfounding variable. In the northern part of the lower peninsula of Michigan, for example, deer habitat production is dependent primarily on the cutting of aspen for the paper pulp market. Deer range quality declines rapidly in the absence of an aspen-cutting program. At the same time, new cuttings retain their value for deer habitat for only a few years. Following cutting, the aspens sprout (nourished by the massive network of roots), grow rapidly, and are above deer reach in two to three years. In a few more years, the closing canopy reduces quality to precutting levels.

The habitat variable can be tied into the deer population model. Since a of the regression of recruitment on posthunt population is a constant (at least within the constraints of a given area), creation of a new habitat can be incorporated into the empirical model by decreasing the slope (*b*) of the regression, and loss of habitat can be incorporated by increasing it. The model can be adjusted easily, but the problem of size of adjustment is a biological problem not easily solved since each case is different. Changes in *b* relative to posthunt population size and population rate of change for the observed George Reserve deer population overshoot are given in figure 11.3. The figure demonstrates the continuity of rate of change in the value of *b*. It also emphasizes the fact that a small overshoot causes a long recovery phase to achieve equilibrium. Although figure 11.3 reflects the general shape, it is only one of a whole family of possible curves. One approach, where new habitat of a known amount is created, would be to express the new habitat as a percentage of the existing habitat and to increase the x-intercept of the regression by a similar amount as an approximation.

Creation of new habitat by the destruction of climax vegetation is a fairly sudden event. Cuttings and fires usually occur over short time periods. For practical purposes, they are instantaneous changes in the slope of the regression model. Assuming a deer population equilibrium prior to the disturbance, a new set of equilibrium values will hold after the disturbance. Furthermore, the possibility of time-lag effects and overshoot are resurrected. Because the state of the vegetation is a dynamic one, succession may be proceeding with a corresponding decline in carrying capacity for deer at the same time the population is increasing. If this were not complicated enough, the size of the deer population determines the feeding pressure on the vegetation, and this is a major variable influencing the rate of vegetation succession. It is theoretically possible to achieve an equilibrium state between vegetation and deer in which succession is halted, with the subsequent deer population at a higher level than it was prior to the creation of new habitat. However, such an equilibrium is difficult to achieve, particularly if rate of succession is rapid. If the deer population is increasing at the same time K carrying capacity is decreasing because of succession, achieving equilibrim at the intercept of these two variables with opposite signs is unlikely, even with management. On a small scale, such situations present almost unsurmountable problems, and population management cannot be very precise. However, a habitat management rationale that stabilizes the creation of new habitat to equal the loss of habitat through succession could, on a larger land scale, achieve equilibrium between deer population and K carrying capacity.

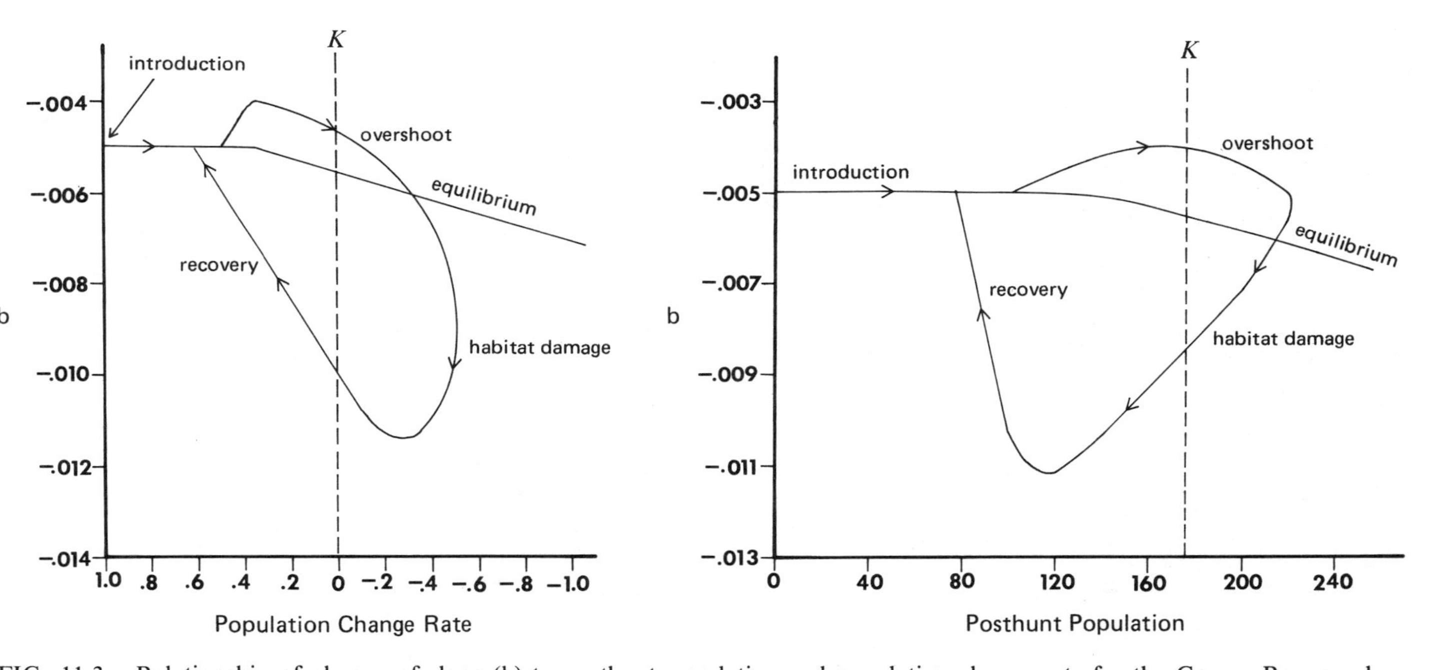

FIG. 11.3. Relationship of change of slope (b) to posthunt population and population change rate for the George Reserve deer population overshoot. Time is shown by the arrows. Since recovery, the equilibrium curve has been followed.

The rate of succession and whether or not the vegetation grows out of deer reach are of major importance in determining the probability of achieving stable equilibria with the creation of new habitat. If succession is rapid, and if the vegetation tends to grow out of reach, time-lag effects will be great. Each situation will be a separate case and must be evaluated as such. Fortunately, the George Reserve is characterized by relatively slow rates of succession, making this study possible. As an aside, it should be noted that problems with these kinds of time lags are less severe when dealing with a climax-adapted ungulate.

CHAPTER 12

Natural Predation and Aboriginal Hunting

Analysis of Wolf Kills

The empirical population model can be used to test the impact of predation by simulation. The primary predator of white-tailed deer in the Great Lakes region was the eastern timber wolf (chap. 1). Selectivity of prey by wolves is different from that of hunters, and this aspect must be explored before the correct impact of wolves can be related to the George Reserve deer population.

Three studies of selectivity of deer by wolves in the Great Lakes region are available: Pimlott et al. (1969) studied wolf kills in Algonquin Park, Ontario; Kolenosky (1972) in the Pakesley area of east-central Ontario; and Mech and Frenzel (1971) in northeastern Minnesota. In each case, the age structure of deer killed by wolves was compared with the age structure of deer killed by humans. It was assumed that the human kill was less selective, hence more representative of the actual structure of the deer herd than the wolf kill. The assumption is safe as far as it goes, but it is necessary to keep in mind that selection by humans also occurs, as witness the selectivity which occurred earlier on the George Reserve even though shooters were instructed to not be selective (fig. 5.1). Any selection will bias the age structure of the kill. This, in turn, biases the calculated living population structure. Still, in the absence of better information, the human-hunter kill will serve as a comparative baseline upon which to evaluate selectivity by wolves.

Deer kill data for wolves and humans for all three geographic areas were converted to l_x series and plotted as survivorship curves for the deer population (figs. 12.1 and 12.2). Note that the formal requirements of the life-table analysis are almost certainly not met, but it is the *relative* shape of the survivorship that is being considered.

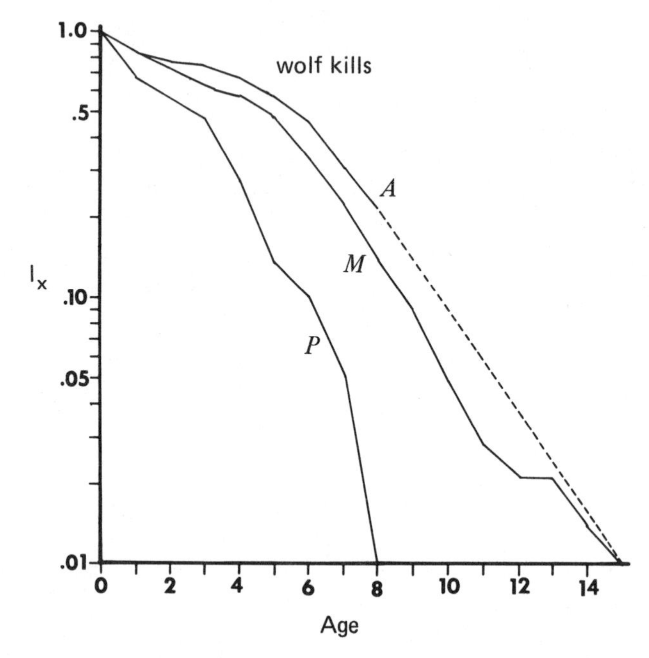

FIG. 12.1. Survivorship curves of white-tailed deer as indicated by wolf predation; *A* is for Algonquin Park, Ontario; *M* is for northeastern Minnesota; and *P* is for Pakesley, Ontario.

It can be seen that human-hunter kills were relatively similar between the three areas (fig. 12.2). The implication is that human hunters showed similar selectivities on the three deer populations. However, the actual age structures of the three deer populations were unknown, and if the age structure in the kill were skewed in the right manner, human selectivity could be masked. The possibility seems rather unlikely, and it seems more reasonable to assume similar human selectivity between the three areas, with the differences in age structure in the kill reflecting relative differences in the actual age structure of the deer populations. The survivorship of deer in Pimlott et al.'s population was greatest, that of Mech and Fenzel slightly lower, and that of Kolenosky the lowest. The conclusion from survivorship curves appears to match the actual situation of the three herds. Deer in Algonquin Park were protected, and the kill statistics for humans were derived from car-killed animals plus those shot for research purposes. Thus, one would expect the deer herd to be closer to K carrying capacity, with a relatively old age structure. The deer herd in northeastern Minne-

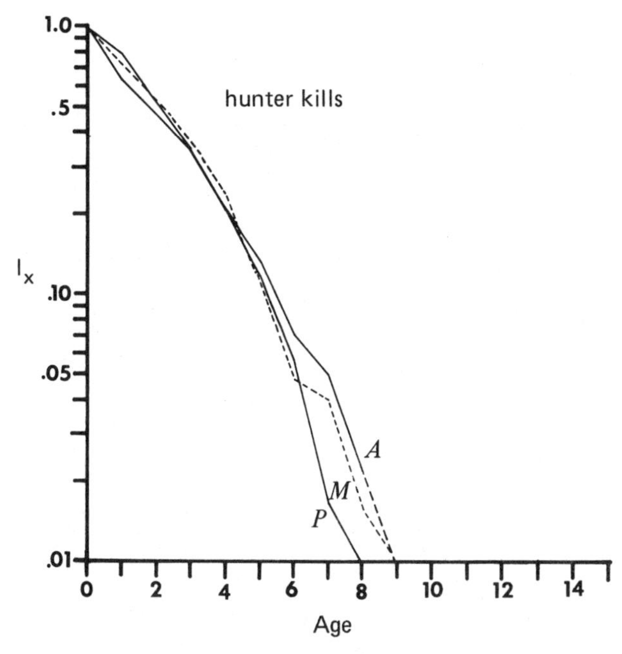

FIG. 12.2. Survivorship curves of white-tailed deer as indicated by human hunting; labels same as in figure 12.1.

sota was subjected to an open season, but it was the belief of Mech and Frenzel that the kill was too light to have much of an effect upon the age structure. Again, if this was correct, one would expect the deer population to be close to K and have an old age structure. The hunter kill in the deer population studied by Kolenosky was relatively high compared to the other two. Kolenosky gives a deer herd estimate in midwinter of 729 in his study area, as determined by pellet-group counts. This census method has many difficulties (Neff, 1968; Ryel, 1971), and the actual population could be substantially different. Nevertheless, it is the only objective estimate of the population size that is available. Hunter kills averaged 83.8 deer per year in the study area, and this number must be added to the midwinter population to obtain the prehunt, fall population of 813 head. Therefore, a kill of 84 would represent 10.33 percent of the prehunt population. In any event, the deer population studied by Kolenosky appears to be substantially lower in relative density than the other two.

The survivorship curves for wolf kills (fig. 12.1) show clear differences in selection of prey by wolves in the three areas. The

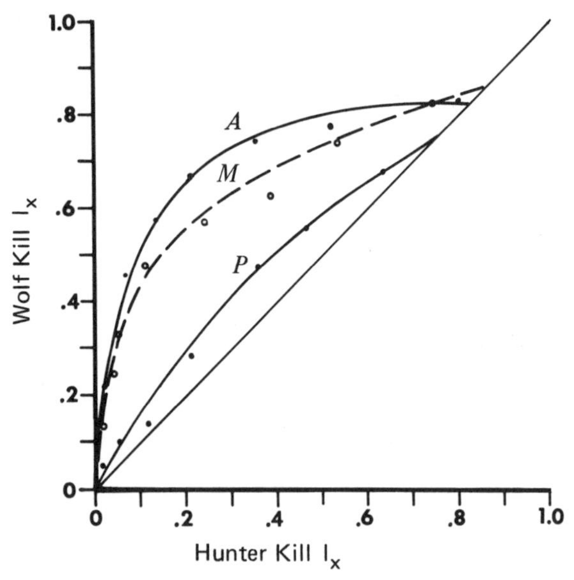

FIG. 12.3. Relationship of l_x of deer populations indicated by wolf and hunter kills for the same area; labels same as in figure 12.1.

wolves of Algonquin Park were highly selective for old animals, as were the wolves of northeastern Minnesota, although to a lesser degree. The wolves studied by Kolenosky were considerably less selective, and the survivorship curve for deer derived from wolf kills approached that for human hunters.

The comparison of selectivity of prey by wolves relative to human hunter kills for the three areas can be most clearly seen by plotting the l_x values given for wolf kills against l_x values for hunter kills taken from the same prey population (fig. 12.3). Once again, Algonquin Park wolves were most selective, followed closely by Minnesota wolves, and with the Pakesley wolves being least selective. These data demonstrate that selectivity of prey by wolves is not constant. It varies according to the density of prey relative to resources and the density of wolves relative to the prey population.

In Algonquin Park, wolves were controlled until 1959. The control was relatively intensive, given that an average of 57.6 wolves per year were removed between 1950 and 1958, and the kill for the last two years was 95 and 62 (Pimlott et al., 1969). Thus, it seems reasonable to suppose that a high deer population and a low wolf population allowed wolves to be highly selective of prey because of the prevalence of old and substandard deer.

A similar situation would seem to have held for northeastern Minnesota. Mech and Frezel believed that human hunting had little effect upon the deer population age structure, which would result in population size being near K carrying capacity (although this population has declined markedly in recent years; see Mech, 1975a, 1977b; Mech and Karns, 1977). Wolves have had some protection in recent years, but strong, antiwolf sentiment among local residents has likely resulted in a sustained wolf mortality (Mech, 1977b). For example, Van Ballenberghe et al., (1975) reported that 73 percent of the known wolf mortalities in their study in northern Minnesota were related to human hunting or trapping.

Assuming that the estimates given by Kolenosky are reasonable approximations, this deer population was subjected to relatively great pressure. Human hunters removed 84 deer from the population of 813 (10.33 percent) and wolves removed an additional 70 (8.61 percent) in the winter between January 11 and March 17. Thus, a minimum of 18.94 percent of the herd was removed by wolves and hunters. If wolves killed deer during the rest of the year at the rate reported for Algonquin Park by Pimlott et al., wolves would kill 133 deer (16.36 percent on a yearly basis) for a total wolf and hunter removal of 26.69 percent of the deer herd. Actually, the summer kill of deer was probably somewhat less, since wolves in the Pakesley area apparently consumed less deer in the summer than wolves in Algonquin (Pimlott et al., 1969; Voigt et al., 1976). Thus, the total estimate would appear to be a maximum estimate. Such exploitation would clearly result in a deer herd with relatively few old or substandard animals. Wolves would be required to work much harder (i.e., to capture less vulnerable prey) in order to survive. Presumably they live at relative densities below what they could attain if human hunting of deer were curtailed.

Unfortunately, comparison of absolute densities of either deer or wolves between the three areas are meaningless without determination of quantity and quality of year-long resources available to the prey population. Furthermore, none of the cases represents a natural balance, since humans have manipulated either deer or wolves (and usually both) and, in many cases, the populations of alternate prey. Despite such difficulties, the studies of wolf-killed deer do show the range in selectivity that is possible under various conditions. The Algonquin Park wolves seem to approach the extreme in selectivity of substandard prey, while the Pakesley wolves seems to approach the limit of wolves to take healthy prey. Presumably, if wolf populations in Algonquin Park were given protection over a longer period, they would increase in number and the system would move in the direction of selectivity as seen in the Pakesley area. It would be unrealistic to expect a stable equilibrium. At this, the northern edge of the range of white-tailed deer, periodic hard winters will inevitably upset the balance, since deer become more vulnerable to wolves in deep snow (Pimlott et al., 1969; Mech et al., 1971). One would anticipate an oscillatory, dynamic balance with the periodicity dependent upon the frequency of occurrence of hard winters.

The range of selectivities by wolves can be examined with respect to the dynamics of the George Reserve deer-population model to assess the impact of natural predators in large game-prey populations. The extreme impact can be estimated by assuming that wolves, in each case, were the sole cause of mortality in their respective prey populations (see chap. 15). This assumption would give an estimate of the *maximum* impact the wolf would have on the prey population. It results in estimates of mortality of 30.57 percent of the prehunt population for the Pakesley wolves, 19.29 percent for the Minnesota wolves, and 17.95 percent for the Algonquin Park wolves. The wolf kills can be treated as removals from the prey population which, over time, would yield average equilibrium values. Since wolves are selective of old animals if they are present in the population, chronic mortality would not be important and the recruitment model can be used to derive the equilibrium population, instead of the population change model. Use of the recruitment also seems appropriate in terms of the wolf data. Obviously wolves kill young, prerecruitment-aged individuals; however, these carcasses are likely to be so completely consumed as to be unrepresented in the kills recorded. A deer must be about six months old before the carcass persists.

Since deer recruitment will shift to balance the removal, the rate of mortality induced by wolves will match the rate of recruitment in the prey population. Thus, the value of the prehunt population (i.e., prewolf predation) can be entered into the rearranged George Reserve recruitment-rate equation (chap. 6)

$$x = \frac{0.9868 - y}{0.005}$$

where y is the wolf-induced finite mortality rate of the deer population and x is the equilibrium posthunt (or postmortality) deer population. Posthunt populations of 136.22, 158.78, and 161.46 are obtained for Pakesley, Minnesota, and Algonquin Park wolves, respectively. Reference to figure 6.6 and table 8.1 will show that even these extreme estimates do not approach the MSY at I carrying capacity (99), and the latter two approach K carrying capacity (176). The true values are almost certainly more towards the K end of the curve, since wolves are not the sole cause of mortality. The estimate derived earlier—that the Pakesley wolves killed 133 deer yearly from a total population of 813—would indicate a removal of 16.36 percent of the deer population. Applying this value to the George Reserve deer model would give a posthunt population of 158.24, or substantially greater than the 136.22 derived by assuming wolves accounted for all mortality. Moreover, the estimate of combined kills of deer by humans and wolves of 26.69 percent of the population would give a posthunt population estimate for the George Reserve of 124.55, or substantially greater than I carrying capacity. Thus, the combined kill is less than MSY and can be supported by the deer population. Although independent estimates of the deer population are not available from the other two areas, it seems reasonable to assume that similar overestimations of the impact of the wolf kill would result.

The Role of the Wolf in the Ecosystem

Natural predators, under equilibrium conditions, are very conservative exploiters of their prey populations. The predator-prey relationship has not evolved to achieve a balance approaching I carrying capacity of the prey population. In fact, the balance is on the K end of the carrying capacity spectrum. The selectivity shown by Minnesota and Algonquin Park wolves is so extreme that it would have a minimal effect on population equilibrium in the George Reserve, even though it appears to be the result of depressed wolf densities. (Selectivities show by the Pakesley wolves are probably more representative of the natural balance.)

Although wolves do not seem to exploit their prey populations at MSY, this does not mean that wolves play an unimportant role in the ecosystem. Selective predation of substandard individuals will serve to eliminate or reduce the frequency of inferior genes in the prey population; but while it is not the author's intent to belittle the culling effect of prey selectivity, it is my belief that this role is overemphasized. For example, antihunting groups frequently cite selectivity of human hunters as not being equivalent to natural predation as a biological justification for the viewpoint. There are several difficulties with this proposition. The first is that intraspecific competition within the prey population is intense, particularly at the high densities which occur in the absence of predation or human hunting pressure. The probability of genetically substandard individuals being successful in reproductive competition is low, and it decreases with increasing population density. Second, the selective pressure imposed by human hunting considerably overlaps that imposed by natural predators; and, in the modern context, it is typically more intense. On Isle Royale, wolves managed to capture only 4.6 percent of the moose they chased (Mech, 1966), and in Minnesota they were observed to catch only one of fourteen deer chased (Mech and Frenzel, 1971). Furthermore, Mech (1970) concluded that the rush is the more important phase of the hunt, with the chase usually being of short duration. Wolves attempted to stalk the prey as closely as possible before beginning the rush. In addition, most of the substandard prey killed by wolves in these studies were inferior due to phenotypical causes (principally injuries and heavy parasite loads) and thus did not represent a threat to the genetic quality of the prey population. It is reasonable to suppose that intraspecific competition in the prey population produces far more substandard individuals than do deleterious genes. And given the strong, intraspecific competition of K-selected species, one can question the likelihood of survival of individuals carrying inferior genes, even in the absence of predators.

Alertness, that capacity to detect and avoid danger, cryptic behavior, and short distance speed are strongly selected under human hunting, overlapping considerably with selection under natural predation. The relatively heavy kill rates of modern hunting constitute a strong winnowing process that could be resulting in genetic changes in the population. One wonders about the successful reoccupation by white-tailed deer of the corn belt over the last twenty years and the extent to which it may be caused by changes in deer behavior. Another and probably more important role of predation is that it tends to stabilize population fluctuations in the prey species. One of the conclusions of this study is that populations at K carrying capacity are inherently unstable, even in the absence of catastrophic events (chap. 9). Natural predation in unhunted populations would be a major force in reducing prey populations from K carrying capacity, thereby dampening fluctuations over time. Furthermore, at low levels of removal, predation would be more effective at dampening oscillations than human hunting because it is selective of old and otherwise substandard individuals. In one sense, it represents "regulation" by predators, but only in the most narrow of interpretations. Full understanding requires knowledge of the nature of the resources of the prey population, both quantitative and qualitative, and their interactions with population density to condition the prevalence of standard individuals. Regulation is a dynamic equilibrium achieved through the linkage of vegetation, prey, and predator.

The equilibrium between deer and wolves is struck on the basis of resilience of the deer population between I and K carrying capacity and the corresponding variation in quality of individual prey. In strongly K-selected species, the equilibrium between predator and prey is established on qualitative characteristics of the prey within the I to K carrying capacity range. It is not due to a numerical response, with the prey being able to reproduce much more rapidly than the predator, as is usually the case in *r*-selected prey species. Indeed, the wolf has a greater reproductive potential than the deer. The average adult female wolf can produce about five offspring per year, as compared to two for deer. Hence, numerical responses could hardly account for coexistence. Nor is it based upon probability of location of prey by the predator due to scarcity and scattered distribution, as is common in *r*-selected species. Presumably, such a mechanism could come into play at some very low prey density, much below I carrying capacity, but its effectiveness for large ungulates is doubtful. The observations of Stenlund (1955) and Kolenosky (1972) suggest that wolves are quite effective in locating prey and use hunting strategies which maximize probability of location. Hypotheses based upon physical refugia would not hold since the two species are about equal in their ability to traverse terrain and cover. Deep snow seems to shift the balance in favor of the wolf (Pimlott et al., 1969; Mech et al., 1971), and running into lakes or streams when pressed by wolves (Mech, 1970; Hoskinson and Mech, 1976) may allow escape by deer that otherwise would have been caught. Still, these variables are far from absolute, and refugia, based upon physical characteristics of the environment, are not sufficient to account for the balance between deer and wolves.

Mech (1977a) has reported that in the zones where the territories of two wolf packs adjoin, deer have higher survivorship, since both packs tend to avoid that area in an effort to reduce the likelihood of aggressive encounters. Aggressive interactions between packs and packs and lone animals is well known (Jordan et al.,

1967; Wolfe and Allen, 1973; Zimen, 1976; Mech, 1977b). The areas around territorial boundaries would function somewhat as refugia for deer, buffering the predator-prey interaction as Mech (1977a) has suggested. Hoskinson and Mech and Mech note that deer in the buffer zones are older, but this does not prove that the deer are less vulnerable to wolves; it may be merely the outcome of wolves not hunting in the buffer zones in the past. However, the effectiveness of such a refugia must certainly be highly variable, dependent on the abundance of deer and the abundance of wolves. Mech (1977a) notes that when deer populations are low, the wolves will be forced to hunt in the border areas where the remaining prey are found and the effect of the refugia may not be measurable. Note that the territorial boundary zones would have resulted in maintenance of the prey population, so food supply would not be limiting on wolves at first. The boundary areas are refugia only during periods of high prey abundance and in themselves would not result in equilibrium between predator and prey. The functional role of the border areas would be to introduce a short, time lag of deer persistence during which intraspecific aggression of wolves might result in a decrease of wolf population. Mech (1977b) reports that intraspecific strife accounted for 25 percent of the adult mortality in the wolf decline in eastern Minnesota. If a balance were to be reached before the deer population was decimated by the combined impact of being hunted by adjoining packs, a dramatic increase in wolf mortality would be required. Moreover, Mech (1975a) reported territorial intrusions by wolves during food shortages. Thus, while such refugia for deer play a role in the dynamics of the predator-prey interaction, persistence of the prey ultimately traces back to the quality of prey. (See Mech and Karns, 1977, for a description of the deer and wolf decline in Minnesota.)

From the early work of Murie (1944) and particularly since the publication of the results of Rabb et al. (1967) on the captive wolves in the Brookfield Zoo where dominant wolves suppressed reproduction by subordinates, great emphasis has been placed on the social regulation of wolf populations. However, interpretation of such results has to be made in a larger context. The Brookfield Zoo is not a natural system. It is axiomatic that social behavior plays a role in population ecology. Competitive relationships between conspecifics are mediated by the behavioral system in nearly all higher vertebrates. Open or scramble competition is seldom observed. Both deer and wolves live in societies and have individual identities. Hierarchical relationships condition the success of reproduction. Individuals ranking high in the hierarchies of both deer and wolves gain access to the best resources, be they food, shelter, cover, or least vulnerable position in the social grouping. Successful reproduction will be higher in these individuals than in those low in the hierarchy.

The question is not whether the behavioral system plays a role, but rather to determine the nature of that role. Is social behavior an ultimate factor? Does the population in question inevitably grow to a given density set by the social system without regard to the availability of food or other resources? This sort of constant trajectory of the population in time and space would deny everything we have learned about ecology and evolution. The inadequacy of such an inflexible concept can be illustrated by a simple example. Suppose one allowed the population to reach the limit set by the social system and then reduced the food base to a level that would support only half as many animals. The social system would determine which of the individuals would be in the half of the population that perished and which in the half that persisted. Should one still insist that social regulation was occurring, or would it not be wiser to view the social system as a proximal factor by which the size of the population is adjusted to the availability of resources in the environment? I propose the latter and suggest that the function of any social system cannot be correctly interpreted without reference to the resources and hazards of the environment in which the population lives and the history of that relationship over time as the social system evolved. Placing animals in an enclosure and removing all of the natural environmental constraints is, by definition, bound to result in a population in which factors other than environmental constraints will set an upper limit. With vertebrates, this will ordinarily be social behavior. The Brookfield Zoo results have been repeated with another enclosed wolf population in Germany (Zimen, 1976), but neither study has bearing on population regulation in the wild state.

In the wild, environmental constraints are present and behavior is a proximal factor in deciding which individuals will survive in the adjustment of population size to environmental constraints. Mech (1975a) has described a situation where a wolf pack declined in the face of a declining deer population, with only the dominant pair surviving. Similar relationships were clearly shown by Zimen's fascinating analysis of the interrelationships of food, population dynamics, and social behavior in captive wolves in Germany. Jordan et al. (1967), Van Ballenberghe and Mech (1975), and Seal et al. (1975) have reported the starvation of wolf pups in the wild. It is concluded, therefore, that social behavior in wolves does not regulate wolf populations independent of the availability of food and other resources. Social regulation is discussed further in chapter 13.

The concept that predator-prey interactions lead to a balance in which both persist has been discussed by numerous authors (e.g., Pimental, 1968; Huffaker, 1958; Slobodkin, 1968; Rosenzweig and MacArthur, 1963; Errington, 1934; and Mech, 1966). Slobodkin coined the term "prudent predator" for this penomenon. It is a catchy term but carries the unfortunate connotation that the predator is showing restraint—a condition that ultimately would call for a group-selection hypothesis. There is no evidence suggesting that the predator-prey balance between wolves and white-tailed deer is anything but the outcome of selection for each to do its very best: for deer to avoid being caught, and for wolves to be the most highly efficient predators possible. Stenlund (1955), Pimlott et al. (1969), and Mech et al. (1971) have shown that given the opportunity, wolves kill deer well beyond what is required to fill their needs. It would appear that, in the ordinary sense of the word, wolves are not prudent and selection would seem to favor the imprudent predator. The argument for prudence based upon the assumption that lack of prudence would lend to extinction of the prey and, subsequently, the predator is specious in the absence of a group-selection hypothesis (Maynard-Smith and Slatkin, 1973).

In a subsequent paper, Slobodkin (1974) argued that group selection was not necessary to explain prudence in predators. He proposed that apparent prudence arose from the predators selecting prey individuals of low reproductive value. These arguments have been reviewed by Mertz and Wade (1976) and Maiorana (1976). Maiorana concluded, and the author concurs, that selecting for prey with a low reproductive value without health also limiting the quantity of prey that the predator can take could not lead to regulation. Refugia could serve this function, but in the absence of refugia, quality of individual prey is the only mechanism that could account for balanced coexistence. Since selection operating on the predator favors killing success, any balance observed must be due to the limitations of the predator to catch healthy prey.

Effect of Aboriginal Hunting on Deer Populations

Only the degree of human impact on white-tailed deer populations has changed. Man, as a hunter, has preyed upon white-tailed deer for thousands of years. What was the impact of aboriginal man upon white-tailed deer populations?

The archaeological evidence from the Great Lakes region is

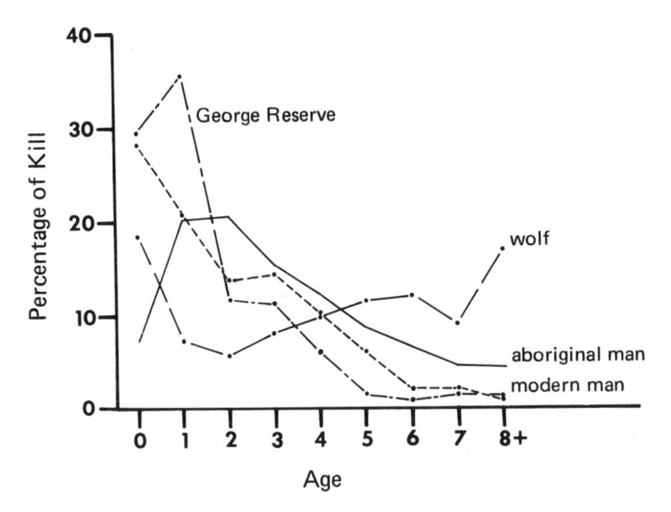

FIG. 12.4. Comparison of age distribution of combined deer kills by wolves, aboriginal man, recent hunters, and the George Reserve

insufficient to answer this question. However, B. D. Smith (1975) and Elder (1965) examined the age structure of a total of 953 deer jaws from eleven archaeological sites in the Mississippi River basin. It is assumed that these results are typical of the eastern deciduous hardwood forest biome, since the age distribution is so similar between the eleven sites studied. McGinnis and Reeves (1958), Guilday (1962), and Guilday et al. (1962) found similar results in Virginia and Pennsylvania.

Age structure for the combined Indian deer kills is given in figure 12.4, where it is compared with the combined wolf kill and modern hunter kill. It is apparent that deer killed by Indians were mainly prime animals, and Indians appear to have been far more selective in this respect than modern hunters, as had been previously concluded by Elder, Guilday, Guilday et al., and B. D. Smith. B. D. Smith has pointed out that there is no way of determining the actual age structure of the population from which the Indian kill was taken. However, treating the data as representative and considering aboriginal man as the sole cause of mortality shows that, by the George Reserve deer model, a population of 134 animals would result. Thus, it is apparent that the maximum possible exploitation was less than MSY. Because the assumption that aboriginals were the sole source of mortality is unlikely to be true, the actual impact was probably substantially less than indicated.

CHAPTER 13

Population Regulation

In the previous chapters, the George Reserve deer population responses have been related to the food base and hunting kill. The existence of a time lag early in the history of the population due to a lack of hunting kill and subsequent deterioration of the vegetation is a further reflection of the relationship of the population to the food base. Heavy hunting kill was the means by which the deer population and food base imbalance was corrected. Chronic mortality and the role of natural predators, as they might differ from human hunting in their effect, was examined. The outcome of the analysis suggests that the population is regulated by a complex of interactions where no single factor accounts for all of the regulation. Even the importance of the ultimate factor—food—will vary depending upon circumstances. But there is a danger in accepting the fit of observations to the explanation as sufficient to demonstrate the correctness of the explanation. The observations may be consistent with other explanations as well. Only by examining alternatives and finding them at odds with, or insufficient to account for, the observations can they be safely ruled out. The purpose of this chapter is to examine such alternative explanations.

Social Regulation

The publication of Wynne-Edwards's (1962) controversial book brought into sharp focus the contrasting viewpoints of two schools of thought on how population regulation functioned. Proponents of the belief that populations regulated themselves suddenly had a champion. Opponents of the idea found it conveniently condensed from a vast and unwieldy literature and restated in a form that could be more readily tested. Prior to Wynne-Edwards, there were many loosely formulated ideas and interpretations interspersed throughout the ecological literature about how populations were regulated. Wynne-Edwards drew the lines. Not the least benefit of Wynne-Edwards's book was that he integrated ecological, behavioral, and evolutionary aspects into his interpretation. The marshaling of evidence, pro and con, accelerated the integration of these fields which had been following largely separate pathways. Most workers now recognize that the separation is artificial; results from one area must be consistent with the others, and ideas in one can be tested by evidence from others.

The most thorough and severe criticism of Wynne-Edwards's hypothesis has occurred in the evolution area, relative to natural-selection theory. The individual or kin level (inclusive fitness of Hamilton, 1964) seems to be the usual, and perhaps only, level at which selection operates. As further tests with empirical data are made, this concept grows more secure. Many problems of assumed altruistic behavior that conflicted with inclusive fitness were created by unwarranted assumptions about behavior. Too often it has been assumed that such behaviors occur at great risk to the giver and benefit to the receivers, when in fact there is little or no evidence to support the assumption. This situation seems likely to continue in view of the ease with which new questions can be posed relative to the difficulty of conducting empirical studies to test them and the current vogue in publishing ideas with the computer (a wonderfully obedient servant that takes the assumptions of the researcher and promptly returns them disguised as conclusions) as the major source of verification. This is not to say that ideas are not important or that computer simulations are not useful. Excitement in a field is generated mainly by ideas. But the rate of exchange between question and answer must be balanced, and empirical tests have been badly lacking in the literature. Be that as it may, there are several models of group selection which appear possible, and it is not safe to put the concept to rest.

Most of the testing of Wynne-Edwards's hypothesis has occurred in the behavioral area. Such observations are relatively easily obtained and can be tested against the predictions and interpretations of selection theory. Few tests via population data have been made, even though Wynne-Edwards emphasized this aspect of his hypothesis. The probable main reason for the lack of testing is the infrequency with which all of the necessary data are available. The George Reserve deer population data would seem to allow this kind of test since both the reproductive effort (embryos) and the success (recruitment) have been determined. If Wynne-

Edwards's hypothesis on population regulation were correct, one would expect that the total effort would approximate the survivorship, since the hypothesis assumes that the number of offspring produced is adjusted to the availability of resources to support them.

Figure 9.6 shows the relationship of reproductive effort (number of embryos) with reproductive success (number of recruits) as population size changes. It is apparent that the attempts at reproduction substantially exceed the number of offspring surviving to approximately six months of age as the population begins to approach K carrying capacity. Clearly, the deer population is not matching the number of attempted offspring to the number the environment will support. These data directly contradict Wynne-Edwards. The data can, however, be interpreted according to an individual selection hypothesis. Each female attempts to maximize her own reproduction without regard to the success of the population as a whole. Given that a few offspring will survive, many females attempt to reproduce. Each female is attempting to reproduce at the physiological maximum of which she is capable.

The physiological maximum has been selected as the best trade-off between the cost of a given reproductive effort and the probability of that effort being successful against the probability of living to further reproductive efforts. Consider the case of the fawn female. At low population densities, fawn females grow rapidly and can reach sexual maturity by fall of their first year. If resources are abundant, the fawn female can maintain her own growth processes *and* conceive and rear an offspring. As density of the population increases, at some point there are not sufficient resources available to the fawn female to maintain her own growth requirements and meet those of producing an offspring. Conception, therefore, would come at the expense of lowering survivorship of the female and reduction of probability of further reproduction at an older age. Since the fawn female, once she reaches recruitment age, has a high probability of further life, the risk of attempting to produce an offspring immediately is greater than the benefit of producing an offspring. In fact, the data show that fawn females only attempt to produce offspring under population densities that allow survival of nearly all offspring.

Cost-benefit trade-offs also occur in older females. If in very poor physiological condition, an individual female may fail to breed in a given year since the added load of pregnancy could result in death. However, the older female has greater latitude in terms of shifting litter sizes between one and three offspring. Furthermore, resorption of fetuses can adjust litter size once gestation is underway, although it appears to be relatively unimportant in white-tailed deer (Verme, 1962; Hesselton and Jackson, 1974; O'Pezio, 1978). Failure to produce milk or giving birth to a small and weakened offspring can discontinue reproductive efforts that are carried to term but are likely to overtax the capability of the female to both survive and rear young. The latter responses allow adjustment of the reproductive effort at a late date, which is necessary because conception occurs long before birth (approximately seven months) and the condition of the female and/or the quality of the environment might well change in the interim. An injury, a particularly hard winter, late spring, or other factors can easily shift the cost-benefit ratio of continuing the pregnancy at the risk of loss of further reproductive efforts.

Verme (1962) has shown that female deer in poor condition gave birth to still born or weakened offspring that died within a few days. He reported that these females produced milk, and Youatt et al. (1965) found that the milk of females in poor condition was of the same quality as that of healthy females. It is significant that Langenau and Lerg (1976) showed that females in poor condition rejected their offspring, including failure to lick the newborn, consume the afterbirth, or allow nursing, even though milk was present in the udder. Offspring from females in poor condition showed care-soliciting behavior equivalent to that of offspring from healthy mothers. However, the females in poor condition showed fearful or aggressive behavior towards the offspring. One offspring was kicked to death by the mother, although most starved to death.

Trivers (1974) has examined the basis for the development of conflict between parent and offspring relative to the proportion of genes they shared in common. He points out that as the offspring grows older, it should be selected to solicit more care than the parent (which is also interested in future production of offspring) should be willing to give. His model assumes that at birth the interests of the parent and the offspring are essentially identical, although he notes that size at birth might be a possible conflict prior to birth.

For K-selected species, particularly large, long-lived mammals, Trivers's model needs to be extended to recognize that the interests of the parent are identical to those of the offspring only at conception, and that many of the goals deviate shortly thereafter. The parent lives in a density-dependent environment, and typically the future quality of resources is not very predictable. Yet, because of the long gestation period, conception must occur long before the specific conditions and major costs of reproduction are known. In deer, for example, the commitment must be made in November for birth to occur the following June, some seven months later. In between these dates lies winter, the severity of which can vary enormously. A good winter may be followed by a dry spring with poor vegetation growth during the period of birth. The occurrence or absence of an acorn crop may shift the energy balance drastically. Obviously the female deer initiates a reproductive effort in the face of considerable uncertainty of the outcome. Still, to maintain her options, she must breed and conceive. As gestation continues, the trade-offs between continuing the pregnancy versus discontinuing it in favor of future efforts must be evaluated. Survivorship of the young to recruitment age is the most reasonable, proximate objective of reproduction. If unfavorable conditions occur, the female may abort or resorb early in pregnancy; or the offspring may be stillborn or allowed to die at, or shortly after, birth by witholding parental care. Thus, the strategies of the female differ substantially from those of the offspring, and since the offspring is entirely dependent upon the female, conflicts are resolved at the expense of the offspring. The situation fits Alexander's (1974) concept of parental manipulation, in which the parent can manipulate the success of given offspring to maximize its own, overall, reproductive success.

Old females seem to reproduce to the end of their life spans. This would be expected, for while young females can opt to forego reproduction in a given year in favor of future reproduction, old females have little expectation of further life. Thus, a reproductive effort foregone may be lost entirely. The outcome is that reproductive effort in the George Reserve deer herd is a function of available resources as expressed through the physiological condition of the females. A similar conclusion was reached for white-tailed deer in Texas by Teer et al. (1965).

Studies of the social behavior of George Reserve deer (Ismond, 1952; Crawford, 1957; Queal, 1962; Hirth, 1977; and Newhouse, 1973) gave no evidence of mechanisms that might be important in population regulation. Deer are not territorial. They occupy overlapping home ranges which are not defended (Queal, 1962). The only time uniform spacing occurs is when females give birth. Nonpregnant females do not show spacing behavior. At birth, a female segregates herself from all other deer, including her own previous offspring (now yearlings). She will not allow any other deer to be near at this time, but the behavior seems to be one more of mutual avoidance than territoriality. Once the new fawn is strong enough to follow at heel, female offspring from the previous year are allowed to rejoin. Yearling males are usually not tolerated, and they join bachelor male groups at this time.

The areas occupied by females following birth are not resource territories. If they were, one would expect to see them established at an earlier time in the spring. Greenup occurs in April, and fawns are not born until June. If resources were being protected, one would expect that territorial behavior would begin at greenup so exclusive rights to an area would be obtained. Additionally, the immediate area of birth is abandoned as soon as the fawn can follow the female, at about four to five days after birth, and the deer then move over an area of eight to ten acres (Queal, 1962). Again, if resources were the point of these exclusive areas, one would expect them to be defended over the high-demand period of lactation. Both Crawford (1957) and Queal concentrated their efforts on the fawning period and found that it was the fawn—not the area—which the female defended.

The function of the exclusive areas at fawn birth seems to be to avoid errors in imprinting, of mother on young and vice versa. If more than one female-offspring pair occurred in a given area, the possibility of incorrect imprinting would exist. Studies of goats and sheep show that females will accept any young if switching occurs shortly after birth (Collias, 1956). New born deer fawns are well-known to be fearless, and they will approach any animal, including man. Once imprinting has occurred, fear responses begin (Salzen, 1962) and the offspring becomes extremely wild. Obviously, the mother has the greatest stake in correct imprinting. If the fawn imprints on the wrong female, on average it will do as well as with the true mother as long as the adopted mother gives complete parental care. In fact, offspring should be selected by attempting to nurse from any female ("thief sucking"). The mother, however, will be investing parental care in an unrelated offspring, to the benefit of another's genes. Therefore, selection strongly favors high discrimination in the females, as evidenced by the aggression shown toward unrelated fawns that attempt to suckle. If females giving birth maintain mutually exclusive areas, there is little chance of error in imprinting and subsequent identification of the correct offspring. Thus, correct imprinting of mother and offspring would seem to be the fundamental reason for pregnant does maintaining mutually exclusive areas at the time of birth.

Both males and females maintain dominance hierarchies (Hirth, 1977). Among males, dominance is extremely important in determining breeding success in the polygynous mating system. However, among both sexes dominance plays a role in resource allocation. During most of the growing season, food on the George Reserve is of relatively high quality and broadly dispersed (fine-grained environment). It is doubtful that hierarchical standing has any bearing upon resource availability in these months for food is much too disperse to be effectively controlled by dominant individuals. However, in the fall food supply is much more variable in quality and patchy in distribution. High-quality green material is distributed in small, favorable sites. Acorns and other seeds and fruits are of very high quality and in poor years are highly localized. Dominance probably becomes quite important at this time, relative to quality of diet. Winter is a time of generally distributed, low-quality foods.

Similar relationships pertain to dominance in relation to favored (i.e., secure) bedding sites and other facets of life history. For the most part, the advantages of dominance relative to resources in white-tailed deer are subtle, but they are general over many aspects of life and cumulative over time. Thus, competition for resources is not completely open, but rather mediated to an extent by the dominance hierarchy, even among fawns. Studies done in pens where insufficient food was available show that dominant fawns survived (Robinson, 1962; Langenau, 1973). This means that shortages are not felt equally by all members of the population; subordinate animals suffer first and most. However, rather than limiting the population in the Wynne-Edwards sense, the dominance order maximizes reproduction under adverse conditions. Animals low in the dominance order are excluded from reproduction by insufficient resources for a successful effort, while those high in the order can continue reproduction successfully because they receive a disproportionate share of the resources. Note the great range in single, twin, and triplet fetuses observed for Y+Ad females on the George Reserve (chap. 4), even at low population density. The social order does not result in the prevention of overexploitation of resources. White-tailed deer have a long history of overpopulation and habitat destruction, a phenomenon also recorded in the early years of the George Reserve herd. Thus, the conclusion about the role of social behavior in deer population regulation on the George Reserve is similar to that reached about wolves in chapter 12.

Critical analysis of the functional role of any behavioral system with reference to population regulation leads, inevitably, to the conclusion that regulation could occur only if group selection were operating. Individual (or kin) selection leads to maximizing reproductive success, although the time of maximization will vary. Consider the bird territorial situation outlined in chapter 10 (fig. 10.3). If one considers the numbers of eggs produced, one could say that the territorial social system limited the population size, since under open competition a greater number of eggs would have been laid. However, the influence of the territorial system at fledging has been to result in maximizing overpopulation. The overall effect of the social system has been to increase the number of individuals that will suffer mortality during the nonbreeding period or to increase the number of individuals that will be nonbreeders in the following year if mortality does not increase.

With a hierarchical social structure such as shown by the George Reserve deer, the outcome is the same. In the absence of a social order, more females would conceive embryos under a given availability of food of a given quality. The idea can be grasped more readily by assuming that the food is not quite sufficient for any female to successfully rear a fawn to six months of age if open competition were operating. Imposing a dominance order would result in some females getting a disproportionate share of the food, and because they did, they would succeed in raising offspring. Those females low in the dominance order would have been eliminated from reproduction at an earlier stage than they would have been under an open competition system. The initial effect of the social system would be to lower the number of embryos conceived, but the ultimate effect would be increase fawn production and further exacerbate the overpopulation problem. A similar outcome would result if some other level of food availability were postulated, say enough to rear ten or thirty fawns under open competition. The imposition of the social structure would always increase the ultimate number of offspring produced, and individual selection would favor the evolution of such a social structure. One could maintain that lowering the initial effort (i.e., the number of eggs laid or embryos implanted) constitutes population regulation, but this would neither be observed over any reasonable time interval for the study of population dynamics nor include the survival of offspring to reproductive age.

In order to lower the ultimate reproductive success, it would be necessary for all individuals to voluntarily give up reproductive effort in an open-competition situation. In a social system, either more than the necessary number of individuals would be required to refrain from breeding or breeders would have to produce fewer offspring than they were capable of, or a combination of both. The social organization of a species is irrelevant to the argument over self-regulation of populations. It is the altruistic behavior of individuals that is the necessary prerequisite for population self-regulation, and this is true whether the species operates under open competition or a social system. The existence of a social system proves nothing about population regulation in that species.

Certainly, social regulation of population size does not exist on a regular basis. There is question as to whether it exists at all in the Wynne-Edwards sense. However, it is also certain that the possibility of group selection does exist, however small, and absolute statements are ill-advised. The burden of proof would seem to lie with the demonstration, in unambiguous fashion, of altruistic behavior as an evolved trait.

There is a school of thought in population ecology referred to as "self-regulation" (Krebs, 1972). It proposes that populations are regulated by intrinsic factors within the population, rather than by factors of the environment. This is not a cohesive school in that the proponents of various hypotheses do not necessarily agree, and other workers use the term without specifying what they mean by it. Some authors have used the term to indicate density-dependent factors, a practice that has led to a great deal of confusion. Others have used it in a sense in which it is ultimately dependent upon group selection to explain its evolution and this would appear to be the descriptive meaning of the term. Restricting usage of the term to this meaning would greatly clarify the arguments, pro and con. Most notable of such hypotheses are that of Wynne-Edwards and the "general adaptive syndrome" hypothesis of Christian (1963; Christian and Davis, 1964). Although the preceding arguments were directed at the Wynne-Edwards hypothesis, similar arguments could be made relative to the latter.

Is "Good" of the Population Selected?

Despite the controvery over group selection, it is still common for biologists to assert that traits evolved for the good of the population or species. To clarify the issue, it seems necessary to specify more precisely just what is meant by the good of the species. In a "might makes right" sense, we might advocate the view that what is good for the individuals is also good for the population. But this is just individual selection in another guise and is not considered relevant to the discussion. Four possible definitions for "good for the species" come to mind: (1) it might be good for the population to possess traits which guard against extinction; (2) it might be good for the population to avoid habitat destruction; (3) it might be good for the population to maximize energy efficiency—e.g., to have the highest conversion rate of energy in plant tissue to biomass of herbivore; (4) it might be good for the population to maximize standing crop. The definitions will be examined in this order.

First, does selection operate to reduce the probability of extinction of a species? At the outset, it can be stated that adaptation is the result of selection over the past history of the species and the environment in which it survived. As long as the future environment is similar to the past environment, one would expect the species to persist. However, if the environment changed drastically, the species might not have enough genetic plasticity to survive; and, if it did survive, the directional selection in the survivors would result in different gene frequencies than before the environmental alteration.

Using mathematical models, Southworth et al. (1974) concluded that the high adaptation of K-selected species resulted in loss of genetic plasticity (i.e., they become canalized), and if habitats change, K-selected have a greater tendency towards extinction than *r*-selected species. This appears to be a valid generalization. However, within K-selected species, not all are equally prone to extinction in the face of changing environments. The degree of specialization, or niche breadth, will determine the susceptability of a species to changes in the environment. Strongly specialized organisms become canalized and cannot respond quickly to rapid environmental change. Generalists are much more adept at coping with environmental alterations. Indeed, the existence of specialized species is a reflection of an evolutionary history where a specific set of adaptions was selected by a highly predictable environment, whereas generalists are the product of fluctuating and unpredictable environments. In this sense, generalists are more "preadapted" to avoid extinction than specialists because they are better at coping with environmental extremes. This is not to say that the probability of extinction is greater in specialists, for the important variable is how rapidly or frequently the specialized environment changes, and many specialists have persisted over a paleontological time scale.

Since selection operates on reproductive success over a time period linked to the life-history parameters of the species in question, the individuals of each generation will do their very best to be successful. The niche constraints within which this success is achieved are set by factors like specialization of structure and behavior and are reinforced by interspecific competition on the "boundaries" of the niche. Individuals within the species compete for successful reproduction and either succeed or fail in this endeavor. The gene pool is the reflection of this intraspecific competition. Viewed in this light, we can say that the imperative of the individual to succeed in passing on its genes is no less, and perhaps even greater in the highly density-dependent situation at high population, if the species is abundant than if the species is verging on the brink of extinction. Reduction in population size does not necessarily favor selection of traits that allow the species to survive, unless they already exist in the population. Given a sudden change in the environment, directional selection will occur in the generalist species. But in the specialist species, the decline of the specialized habitat with environmental change will select for the most highly specialized individuals in the population and set the species on a collision course with extinction.

The conclusion about whether or not species possess traits that guard against extinction, therefore, is that they do not. The adaptations of a species are a reflection of its past evolutionary history, and its future outlook is dependent upon the degree to which the future course of events resembles the past. Highly specialized species are most vulnerable to changing environments, and this is a particular problem with the immense alterations of the natural world produced by technological man. A larger number of threatened and endangered species are specialists whose particular habitat is disappearing rapidly.

The second possibility is that it would be good for a species to avoid destruction of its own habitat. Do species possess such traits? Because nondestruction of habitat results from achievement of a sustainable equilibrium, this trait is a characteristic of K-selected species and should be found in these populations if it exists.

Southworth et al. concluded that in animals with permanent habitat, overshoot—with destruction of habitat—would be selected against. However, their model considered the population as a unit, implying that group selection was operating. An individual selection model leads to a considerably different conclusion, as previously discussed. Deer are K-selected and have a notable history of habitat destruction. They are not, however, as K-selected as a number of ungulates which occupy climax situations, and the periodicity and amplitude of habitat destruction in these species is less than with deer. Still, from elephants and hippopatami in Africa to caribou in the arctic, long-term buildups and declines seem to be the pattern, rather than a stable equilibrium. It is true that man has altered the balance by influencing vegetation and predator abundance, but if nondestruction of the habitat is a species characteristic, shouldn't it operate independently of these other forces? And even when man has had little impact, environmental variability and time-lag effects have resulted in oscillations; dynamic equilibrium is achieved only over long time periods. From both theory and empirical evidence it seems that even highly K-selected species do not possess traits to avoid habitat destruction. Indeed, only in those situations where the physical environment is extremely stable and predictable, or where man intelligently assumes the role of regulator, would one expect to see a stable balance.

It is concluded, therefore, that populations are not selected to avoid habitat destruction. At first thought the conclusion may seem to be at variance with the discussion of equilibrium tendencies based upon L values in chapters 8 and 10, but it is not. Recall that L values are derived from the net reproductive rate (R_0) of the life table, and that both reproduction and survival schedules are incorporated in this statistic. Thus, the tendency to track K carrying capacity is expressed through the total dynamic responses of the population. The question here is if the control of reproduction and mortality schedules is vested in some vague and poorly defined "wisdom" of the species, or is it a product of the environment? K-selected species have low reproductive rates, clearly related to the lessening of the likelihood and degree of overshoot. Some would consider this fact as demonstration that the species is selected to avoid overshoot. But the reproductive rate has been selected over time by those individual females producing the most *surviving* offspring, being the ones who pass on genes to subsequent generations. It has been a tenet of population ecology since Malthus that overpopulation of offspring is a universal trait among animals, including even the most highly K-selected species. The tendency to produce excessive numbers of offspring despite a long history of being K-selected is readily explained by individual selection. It would be more difficult to explain the lack of balance between number of offspring and available resources if group selection were operating. It is the environment that determines which and how many of the excess individuals will survive to reproduce, and this selective process is repeated with each reproductive period. Reproductive rates observed today are the reflection of a past history of environmental selection—a selection similar to that occurring today. Historically and currently, the environment has been the ultimate source of population regulation.

The third definition of good for the species involves the question of whether a species maximizes energy-transfer efficiency. Energy-transfer efficiency can be expressed in many different ways (Odum, 1971). The two of interest here are the ratio of standing crop biomass of plants to standing crop biomass of herbivores (trophic-level production efficiency), and the ratio of food intake to production of new biomass of the herbivore (utilization efficiency). If selection favored utilization efficiency, it would require a lowering of the population size below that which the environment would support, since greatest utilization efficiency is realized at I carrying capacity. If verified, it would suggest the operation of group selection. Since the natural tendency of the population is to grow to K rather than to stabilize at I carrying capacity, the possibility can be ruled out. Utilization efficiency does not seem to be an evolved trait of the species.

However, trophic-level production efficiency *would* be approximated, because at K carrying capacity, the highest standing crop of herbivore biomass would be achieved. Because of intense, intraspecific competition, plant biomass intake would be maximized. The effect would be to lower the standing crop biomass of plants, thereby increasing the ratio further—a result also correlated with the tendency of the herbivore population to overshoot K carrying capacity.

Maximization of trophic-level population efficiency is approximately consistent with the outcome one would expect from an individual selection (inclusion fitness) hypothesis, but at least one discrepancy argues against the conclusion that selection occurs at the population level which maximizes trophic-level production efficiency. The discrepancy is that the energetic cost of aggressive social behavior (MacArthur, 1972) is greatest in the crowded conditions of a population at K carrying capacity. Most K-selected species spend a considerable amount of energy establishing and maintaining dominance position, territory, etc. If selection were working on the species level to produce the maximum standing crop, it would be obtained by having a completely cooperative, noncompetitive social organization in which the species would not have to spend energy in competitive, social relationships. Yet, the very essence of a K-selected species is the competitive ability of individuals. Note that social behavior, in this context, does not protect the environment. The impact on the environment of a given removal of plant biomass is the same, whether it is used to maintain a high biomass of cooperating herbivores or a lower biomass of competing ones, because the latter use part of their energy in social competition to secure resources and reproductive advantage over competitors.

The overall conclusion, therefore, is that a rigorous examination of population traits will demonstrate that selection does not operate for the good of the species or population. This result, based on population dynamics, is consistent with the outcome of individual versus group selection examined in the light of genetics and natural-selection theory. These are fine distinctions, but they are certainly not trivial. They are fundamental to understanding population regulation.

Dispersal and Population Regulation

In addition to reproduction and mortality, immigration and emigration can produce population changes in a given area. In white-tailed deer, which do best in disturbed areas, the ability to disperse is well developed. It is mainly the yearlings that move away from their area of birth. Hawkins and Klimstra (1970) found that mainly yearling males dispersed (about 80 percent of the yearling males in the population) from Crab Orchard National Wildlife Refuge in Illinois, and that such dispersal was an important factor in reducing the size of the population on the refuge. Similar results were reported by Kammermeyer and Marchinton (1976).

With the exception of a few escapees prior to 1963, when the fence was increased in height, movement into or out of the George Reserve has been absent. This is probably the most unnatural aspect of the deer population. How might normal dispersal be related to the present studies?

To evaluate dispersal one needs to know the variation in density over space in a given species. One also needs to know the relationship of that density to carrying capacity, for as demonstrated for mule deer in Utah by Robinette (1966), dispersal may occur to areas of greater density if carrying capacity of the environment is correspondingly higher. If it were assumed that the George Reserve was a small area in a much larger, homogeneous environment with similar deer population densities, one would expect that without a fence as many animals would immigrate as emigrate. There would be no change in population size, even though there was an exchange of mainly young individuals. This is the hypothetical assumption under which the present analysis was made. Dispersal as a population phenomenon cannot be studied on the George Reserve, although the absence of this variable has improved the control of the experiment.

In evaluating dispersal as a population-regulating factor (e.g., Lidicker, 1962), it is necessary to include a natural area over which population exchange might occur. The arbitrary selection of a study area that is less than a complete natural unit can lead to conclusions that are artifacts of the definition of the study area. Assume that a block of favorable deer habitat was surrounded by a larger area of much poorer habitat. Recruitment would be relatively high in the favorable habitat as compared to the surrounding area, and net emigration would be to the surrounding area. Dispersal would be mainly by young individuals unable to compete with more dominant ones for resources in the favorable habitat. This is the situation reported by Hawkins and Klimstra, and Kammermeyer and Marchinton. With reference to the central core only, a conclusion of self-regulation through dispersal of socially subordinate individuals would be reasonable. But such a conclusion would be artificial relative to the total natural unit, since it ignores the fate of dispersing animals. It assumes, implicitly, that the outside world has an infinite capacity to absorb dispersers. This is patently not the case. If net dispersal was continually outward and the surrounding area continued to support low densities, it would be apparent that the recruitment rate *plus* the addition of dispersers could not raise the population size. The mortality rate would have to be high, caused by such things as lack of food, winter cover, greater hunting pressure or vulnerability to hunting, and dogs or natural predators. The point is that animals would die because of environmental constraints. Thus, in the good habitat, reproduction exceeds mortality and balance is achieved by dispersal. In the poor habitat, reproduction plus immigration is balanced by mortality. For the total area, reproduction, which is disproportionately concentrated in the good area, is balanced by mortality, which is disproportionately concentrated in the poor area. Ultimately, the constraints of the total environment set the limit to total population size. Both social behavior and dispersal play an important role in the process, but the apparent regulation by dispersal is an artifact of considering the central core as a complete unit and ignoring the fate of individuals that disperse. Again, it is necessary to evaluate total social behavior with reference to environmental constraints.

Although a functional outcome of dispersal is the relief of population density on the core area, it does not speak to the natural selection process leading to the evolution of dispersal. Dispersal to reduce density of the population could only have evolved if group selection were operating. Assuming individual selection underlies the mechanism, if more than 50 percent of yearling males disperse over time, the average reproductive success of dispersing males must exceed the average success of nondispersers. In a population near K carrying capacity, the average success of the nondispersers will be low because the probability of surviving to dominant status is slight. Given the patchy and unpredictable creation of favorable habitat in a subclimax species such as the white-tailed deer, dispersers have a highly variable success. Most fail to find favorable habitat and probably do poorly. But a few do locate favorable habitat and they have high reproductive success, thereby raising the average success of dispersers above that of nondispersers. Consequently, the evolution of dispersal by individual selection is easily accounted for. Conversely, climax species are often very poor dispersers (e.g., Geist, 1971), and this is what would be expected from individual selection.

CHAPTER 14

Natural Selection and the Sexes

Niche Separation between the Sexes

The results of recruitment in relation to sex (chap. 6) suggested the possibility that males and females were not competing directly. Sexual dimorphism in birds relative to niche separation has been discussed by Selander (1966). The hypothesis proposes that different morphology between the sexes allows them to exploit different feeding niches, thereby reducing the amount of competition between the pair and increasing the total food available to be directed toward reproduction. Sexual selection, presumably, was the origin of these differences in morphology; but once they appeared, competition reduction became a selective force on feeding behavior or structures, with a resultant differential niche by sex. Differential resource use by sex has even been proposed for plants (Freeman et al., 1976).

The same arguments could be applied to mammals, but little work on the subject has been done. Pair territories occur in small ungulates living in dense forest undergrowth (Estes, 1974; Jarman, 1974), but these species show the least sexual dimorphism and there is little reason to suspect that niche separation by sex occurs. However, nonterritorial ungulates in forest habitats, and all ungulates in openlands are polygynous and show strong sexual dimorphism. There is a relationship between amount of sexual dimorphism and degree of polygyny, as would be expected from the theory of sexual selection. Few polygynous ungulate males contribute directly to the parental care of offspring. A few exceptions occur, e.g., plains zebras (*Equus quagga*) (Klingel, 1969), horses (*Equus caballus*) (Feist and McCullough, 1975, 1976) and vicuña (*Vicugna vicugna*) (Koford, 1957; Franklin, 1974). These species show yearlong harem groups with a single dominant male, where parentage is assured. In most polygynous ungulates, male parentage is uncertain, and males do not defend or lead the female-young groups in times of danger. In many polygynous ungulates, males live in separate groups and areas from females except during the breeding season. This is true of at least some other polygynous mammals as well. For example, Kaufmann (1962) has shown that male coatis (*Nasua narica*) occupy separate ranges from females. Smythe (1970) has suggested that food habits are different between the sexes. Thus, both differences in food selection and spatial separation could result in niche differentiation between the sexes.

While spatial separation is pronounced in ungulates, differences between the sexes in food habits are difficult to demonstrate. On the George Reserve, rumen samples examined for food habits show some differences but they are so slight as to be possibly due to sampling error. Nor do other characteristics of the rumen material suggest pronounced differences in feeding behavior between the sexes, although the differences might be subtle and extremely difficult to measure. For example, the size of the rumen in relation to body size will determine the quality of forage that can be digested, other things being equal (Short, 1963); and the larger the body size, the lower the basal metabolic rate per unit of body weight (Kleiber, 1961). Therefore, males, with their larger size, could do equally as well as the smaller females on a lower average quality of diet. On the George Reserve, males have a higher rumen fill (on a dry-weight basis) relative to body size than females. This could be interpreted in several ways, including one favorable to the niche-separation hypothesis; i.e., males consume a greater amount of lower quality forage. However, the results are ambiguous, and I am not aware of any clear demonstration of a difference in food habits between the sexes in ungulates.

In spatial separation of sexes, outside of rut, the quality of habitat used by the sexes in their respective areas is seldom clearly determined. Although there are exceptions, it is my impression from the literature and personal experience that males of most species usually move into more marginal habitat during the nonrutting period. However, their density is also usually lower, and the relationship to resources is not clear. Again, the diferences are so subtle as to defy measurement, but on a cumulative basis, they could be extremely important. In some species the separation is very strong, virtually without overlap. On the George Reserve, separation is much less complete and fairly subtle. As noted by

Hirth (1977) and Newhouse (1973) and as demonstrated by my own data, males and females are seldom found together in mixed groups except during the rut. The data alone do not demonstrate a spatial separation, however, only different social units.

Between the beginning of July, 1969, and the end of June, 1970, forty-seven night-spotlight samples were conducted over a fixed vehicle route covering all of the reserve that can be observed from a vehicle. Figure 14.1 shows the distribution of females and fawns during the nonbreeding and breeding periods. Breeding periods included the time over which courtship of females by males was observed by Hirth. Areas favored by females between the two periods appear unchanged. During the breeding season, males approximate a random scatter over the area. During the nonbreeding period, however, the distribution of males is more clumped. Overlap with female and young groups is still fairly great in the areas favored by males, but males seldom are observed in the most favored female areas in the central part of the reserve. Thus, while spatial separation is relatively poorly developed on the reserve, the tendency appears to be present, and perhaps it cannot be fully expressed in an enclosed area of this size.

Why do the sexes separate? It does not seem to be due to social antagonism. Occasionally the sexes will be observed intermingling in both areas, with no apparent conflict. Certainly the females cannot drive the more dominant males from the prime habitat into the marginal areas (although females do drive males away from young fawns). And typically it is the males that move to the female areas during the rut and away after the rut. Two questions emerge: (1) Why don't females also move into the male areas? and (2) Why don't the males remain with the females?

One apparent reason for the females not moving into male areas is that the danger of predation is frequently greater there, and the young that follow the females are the most vulnerable. Because of large size and strength, males may be less vulnerable than females and young to predation. Yet, the evidence for higher predation on males looks convincing. In white-tailed deer killed by wolves, males, particularly adults, predominate (Pimlott et al., 1969; Kolenosky, 1972; Mech and Frenzel, 1971), even though adult females are more common in the population. Similarly, adult males were most common in mule deer kills by mountain lions (Robinette et al., 1959; Hornocker, 1970). Apparently the greater size and strength of males is offset by the risks associated with sexual competition. Males are more likely to incur injury, and in

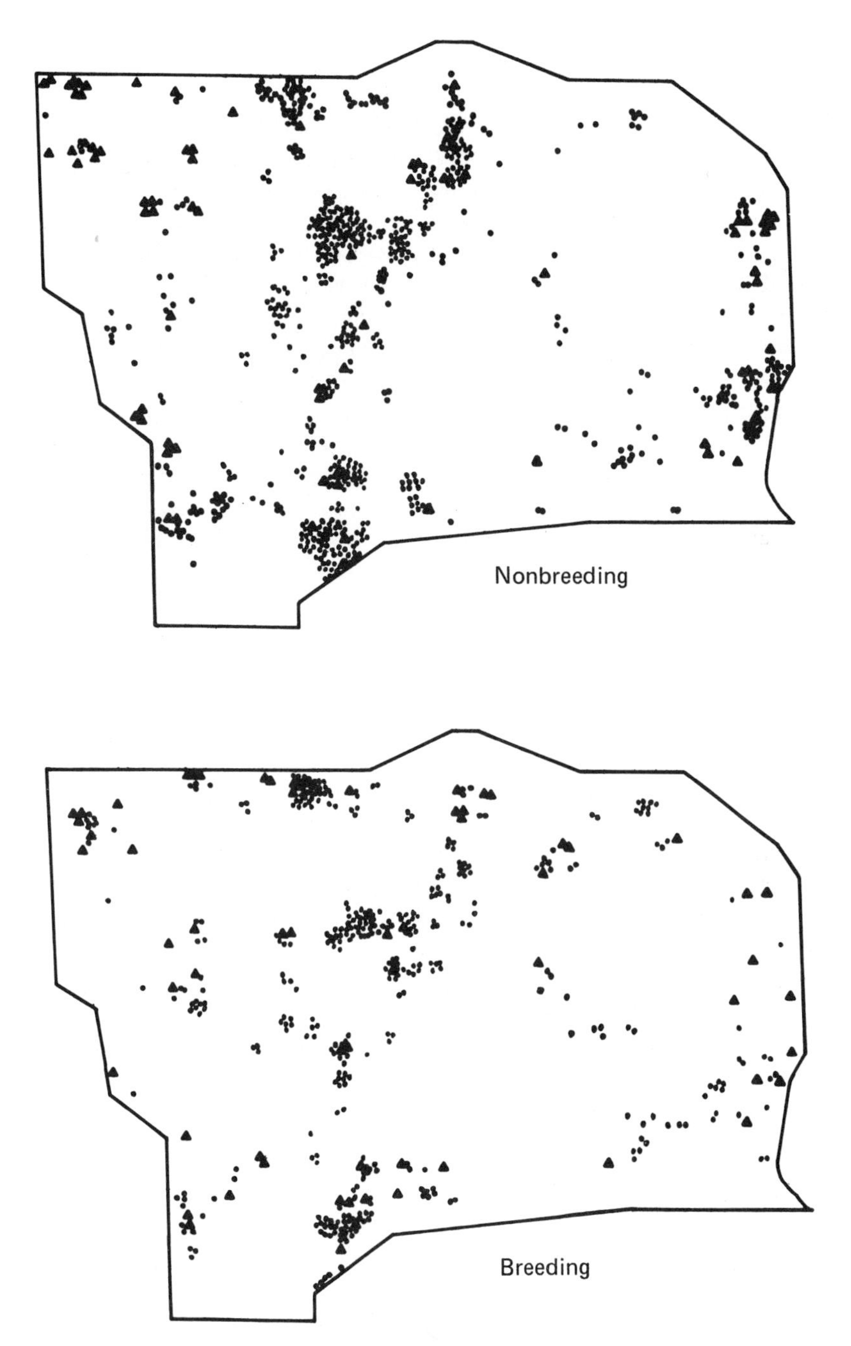

FIG. 14.1. Distributions of males (triangles) and females (circles) during the nonbreeding (Jan.–Sept.) and breeding (Oct.–Dec.) seasons on the George Reserve

many species their physical condition deteriorates during rut and they may be more vulnerable at this time. Females may avoid marginal areas to reduce the likelihood of predation.

Greater predation could account for females not moving into male areas, but why would males leave female areas to move into marginal habitat? The move may benefit males by (1) increasing their survivorship, thereby allowing increased future reproductive effort, or (2) increasing the survivorship of their offspring and (3) increasing the survivorship of females as potential future mates. Since the evidence suggests that survivorship of males is less than females and predation is heavier, increased survivorship could not be the benefit of separation from the female groups. Hence, the increased survivorship of females—as future mates and contributors of parental care to offspring carrying genes of the dominant males—seems to be the apparent benefit to the males. It would further seem that absence of males could only contribute to the survivorship of young (in species in which males do not defend young), through reduction of male competition with reproducing females and immature offspring for common resources. This is the reason proposed by Geist and Petocz (1977) for spatial and habitat segregation of the sexes in bighorn sheep.

The benefit of increased reproductive success must be great enough to offset the survivorship cost of moving into marginal areas. But in polygynous ungulates, only the most dominant males engage in successful breeding. Since they account for nearly all of the offspring, reducing competition for females and young will increase their own fitness. There is little difficulty explaining the selective advantage of the most dominant males moving from the best habitat for females and young, as Geist and Petocz have suggested. But why would less dominant males, who have few or no offspring (i.e., no benefits) assume the cost of moving from the best habitat? Clearly, they cannot have the same benefits as the most dominant males. The answer may lie in the nature of sexual competition between polygynous males. A young male can only hope to become a dominant male by engaging in hierarchical competition over a long time period. The achievement of dominant status by young males requires outliving older, stronger males and dominating males of similar age. It is not enough to maintain position; to succeed, the young male must continually strive to move up in position. This requires that he associate not only with animals of lower rank, but also with those of higher rank. Only the most dominant males do not have need of moving up in rank, and in many species, they often become solitary in the nonbreeding period. Success, as in any other contest of strength, skill, and endurance, comes from long and diligent training. A young male, choosing not to join all male groups, would not be able to obtain the necessary skills to compete successfully. Therefore, the young male has no choice but to become associated with other males. Consequently, if the most dominant males separate from the best female and young habitat to increase their own reproductive success, less dominant males must do likewise. Striving for greater hierarchical position is important for all males in the hierarchical scale.

On the George Reserve, yearling males become independent of the mother when she drives them off prior to parturition in May or June. The yearling males wander about, alone or with other yearling males. During this time, they have high interaction rates with each other and with older males in buck groups (Hirth, 1977). By late summer, social relationships have become more settled and interaction rates decline for all males. Yearlings in all-male groups are persistently threatened, yet they do not leave the group. There is little permanence in the male groups—males change associations frequently, as one would expect from the need to interact with animals higher in the hierarchy and the small group size on the reserve. Mean group size for all male groups is less than two (Hirth, 1977).

This explanation would account for the formation of all-male groups and the spatial separation in those species in which it occurs. Much more work needs to be done on the question of niche separation by sex in mammals. Hopefully, formulation of this hypothesis will stimulate the collection of data needed to test its validity.

Selection of Sex Ratio

The finding that sex ratio varied with the density of the prehunt population (chap. 4), with males predominate at high population, is of considerable interest. Primary sex ratios unbalanced in favor of males are typical (Severinghaus and Cheatum, 1956; Taber, 1953) and can be explained easily on the basis that most deer populations have been lightly exploited and live at relatively high densities. The variation in sex ratio would not be discovered except in circumstances such as the George Reserve studies or penned studies, such as Verme's (1965). The conclusion that it was total population, rather than one or the other sex, that was responsible for the variation in sex ratio of offspring is expected, given individual selection. If offspring of a given sex have higher fitness, it would be to the advantage of both parents to invest more effort in the production of that sex. Thus, the physiological mechanism responsible for differential sex ratio in offspring might be vested in the male or female parent, or both since their strategies relative to sex of offspring would be the same. Some workers might suggest that differential sex ratio in offspring is a fortuitous outcome of a physiological mechanism. This would be begging the question. Since the proximal control (the physiological mechanism) is also subject to the forces of selection, one must still account for the ultimate benefit of higher fitness in explaining its evolution. A "how" answer will not suffice for a "why" question. The selective advantage of varying sex ratio dependent upon population density still must be sought to answer the "why" question.

At this point, a caution should be noted. The previous arguments and analysis apply to the "mean" individual and not to the actual individuals in the population. Dominance position might have a pronounced influence in terms of sex-ratio selection, as will be discussed, but since specific identity and hierarchical positions were unknown in most cases, the quantative analysis pertains only to the mean individual. This caution also applies to general theories of sex-ratio selection, to which we now turn.

Verme (1965) hypothesized that under good food conditions, sex ratio was selected in favor of females to allow more rapid population growth. This would be a fortuitous outcome, but it could not be the process by which it was selected unless group selection were invoked as an explanation. It is still necessary to explain the individual advantage, since what is good for the population will evolve only if it is advantageous to the individuals, and it is under this principle that selection operates.

Downing (1965) suggested that the first matings of males tended to produce predominantly female offspring, while later matings produced mainly male offspring. Obviously this could be a proximal mechanism, but it would still demand an ultimate evolutionary explanation to account for its selection. Even as a proximal mechanism, sex differential by mating order of a given male is not a very convincing idea. Numerous studies in recent years of polygynous ungulates conclude that most breeding is performed by dominant males (e.g., McCullough, 1969; Geist, 1971; Kitchen,

1974; Coblentz, 1974; Feist and McCullough, 1976). Hirth (1977) has shown the same to be true of white-tailed deer. In an enclosed population such as Downing observed, it is likely that all of the breeding was done by one or two males.

Robinette et al., (1955) reported that mule deer females having their first offspring have a higher proportion of males (122:100 females) than older females (106:100 females), and McDowell (1959) reported similar results for white-tailed deer. Since the proportion of young females in the population increases as density decreases, age of females could be an important factor. However, the relationship reported by these workers would mean that the greater number of young primiparous females at lower density should give a preponderance of males in the offspring, when the reverse, a preponderance of females was observed. Again, the explanation could not account for the results obtained.

Trivers and Willard (1973) have argued that females in good condition should produce a disproportionate number of males, while females in poor condition should produce a preponderance of females. This is based upon the expectation that females in good condition will have superior offspring and, in polygynous species, that superior males will have disproportionate success in contributing genes to subsequent generations. On the other hand, nearly all females reproduce. Therefore, females in poor condition are unlikely to have male offspring that can compete successfully with male offspring from females in good condition. These females should maximize their reproductive success by producing female offspring. On a gross population basis, the hypothesis does not hold. At low density, when most females are in good condition on the George Reserve, a preponderance of female offspring resulted. But Trivers and Willard probably never intended condition to be expressed on an absolute scale. Even at low densities, some females are larger or in better condition than others, and the same is true at any given density. Thus, a critical test of the hypothesis requires data on hierarchical position of individual females, the sex of their offspring, and success over the lifespan of the offspring. Such data are not available for the George Reserve deer population; however, the disproportionate number of males produced by females having their first offspring would argue against the hypothesis in deer.

Fisher (1958) proposed a hypothesis to account for selection of the primary sex ratio. He concluded that because each sex contributes 50 percent of the genes to the subsequent generation, parents would be selected to give equal amounts of parental investment to each sex. The total parental investment in producing males should exactly equal the total investment in producing females. If there is a differential mortality among offspring of one sex, say males, the cost of each male reared would be greater than the cost of each female reared, while the cost of each male born would be less than that of each female born. Therefore, if the total cost of producing males was less than that of producing females, natural selection should shift the primary sex ratio in favor of males until the cost of each sex was equalized. Fisher's "principle," as it has become known, has been explored in greater detail by numerous authors (e.g., Kolman, 1960; Verner, 1965; Hamilton, 1967; Leigh, 1970). In general, their studies have supported the principle.

The white-tailed deer seems to meet the predictions of Fisher's principle. Typically, males predominate in the primary sex ratios, and they have a higher mortality rate than females during the period of parental care. The decline in proportion of male offspring with decreasing density in the deer population would also seem to conform. If all offspring conceived were to survive, then the total investment of parents in males would greatly exceed that for females because the cost of producing males is greater. Male offspring are larger and have a higher metabolic rate than females. Natural selection should return the investment in each sex to equality by decreasing the proportion of males in the primary sex ratio.

In deer, there is no reason to presume that the sex ratio relative to density is genetically fixed. Although this study involved shooting to change the relationship of population density to resources (and unknowingly, genotypes may have been differentially removed), this is not the usual case in nature. Usually, the mechanism of change of population density to resources results from the sudden and drastic creation of suitable habitat, e.g., fire and storm—the typical circumstances under which sex ratio in deer was selected. Because of the relatively slow rate of increase in K-selected species, the fact that most offspring survive and males (which because of polygyny could greatly influence gene frequencies) do not breed until they are quite old precludes the idea that shifts in gene frequencies could account for the shift in sex ratio, just as shifts could not account for the results in Verme's feeding trials where females were assigned to diet at random. The conclusion, therefore, is that the shift in primary sex ratio is facultative and somehow expressed through the phenotype. Williams (1966)

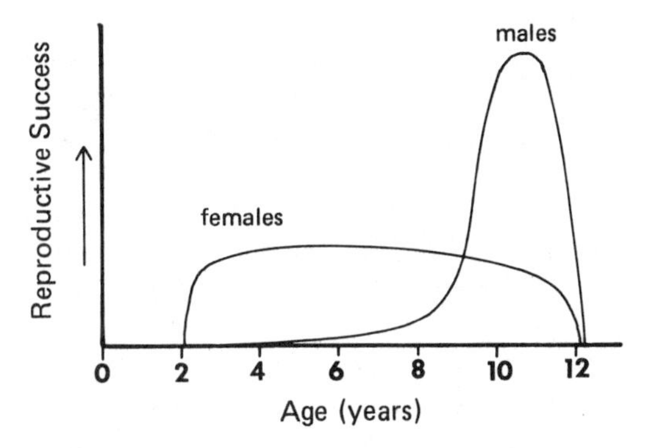

FIG. 14.2. Generalized age distribution of reproductive success of males and females for the George Reserve deer population at a population size near K. Areas under the two curves are equal.

has explored the evolution of facultative versus fixed responses. A fluctuating environment would favor selection for a facultative response if the value of the sexes varied with the changes in the environment. Deer, as subclimax adapted animals, live in greatly fluctuating environments.

Consider the reproductive strategy of the individuals in the population. Each individual attempts to maximize its contribution of genes to subsequent generations, and sex of the offspring is one of the variables. Offspring, once conceived, have a specified sex, so selection for sex ratio must be expressed at the time of conception. On a long-term basis, Fisher's principle would hold, and the best strategy would be to conceive slightly more males than females in deer. But the differential time of reproduction between the sexes, in conjunction with periodic and abrupt creation of favorable habitat in a subclimax adapted species, can shift the relative values of the sexes in a way that will select for differential parental investment in the sexes and sex ratios which deviate from the Fisher equilibrium ratio. At population equilibrium at K carrying capacity, the reproductive contribution of males will equal that of females, although the time over which reproduction occurs will be different for the sexes (fig. 14.2). It is this difference in timing that shifts the values of males versus females over short time periods. The advantage of females over males occurs during periods of rapid population growth. This can be seen from the greater yields (i.e., greater recruitment of young) caused by sex ratios unbalanced in favor of females at low population densities (fig. 9.5).

To illustrate, assume that the George Reserve deer population is at equilibrium at K carrying capacity over a long period of time. Most conceived offspring die before recruitment age. Female survivors of a given year-class reach sexual maturity relatively late (1.5 or 2.5 years) and produce small litters that have low survivorship to recruitment age. Fathers of these offspring are from year-class cohorts of a much earlier time. Few offspring of the females of a given year-class are fathered by males from the same year-class. Surviving males of the year-class produce no offspring until late in life, when a few survivors reach dominant status and account for most of the offspring. Most of these offspring are from mothers of year-classes born later than that of the fathers. As long as the equilibrium persists, sex-ratio selection will favor the Fisher ratio (an evolutionary equilibrium), and 50 percent of the genes will be contributed by the females and 50 percent by the males of the year-class in question.

Deer, however, are adapted to subclimax vegetation, and their success depends upon the periodic destruction of mature vegetation by fire or similar events. So, rather than having been selected in a relatively constant environment, deer have evolved adaptations to periodic, abrupt shifts in K carrying capacity. Under climax conditions, K carrying capacity is extremely low, and following disturbances of the climax, extremely high. Total increase in carrying capacity is determined by the extent of favorable habitat created by fire or other causes. Such a fluctuation in K carrying capacity could be produced by a massive, habitat-management program introduced suddenly on the George Reserve. Assume such efforts resulted in a doubling of K carrying capacity. The change can be simulated with the empirical model by increasing the slope (*b*) of the recruitment rate and population change regressions (fig. 7.5) so that the *x*-intercept is twice what it was. Because of the different times of expression of reproductive success between males and females, males in year-classes reaching dominance at the time of doubling the habitat have a disproportionate reproductive success, leaving more genes than females of the same year-class (figs. 14.3 and 14.4). The total offspring production by all male year-classes born prior to doubling of habitat and influenced by the change in habitat (i.e., back to year-classes where male and female success was equal) is 552; that of females is 470. Thus, the differential advantage of males would favor a sex ratio of 1.17 males:1 female (54 percent male). Neither sex would be favored in year-classes after the doubling of habitat, because the advantage of

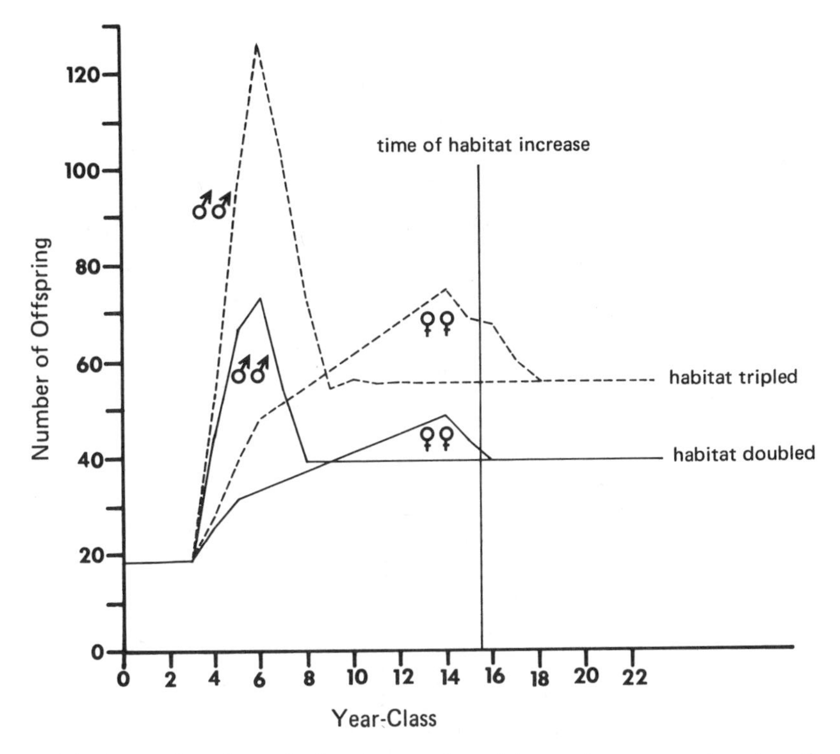

FIG. 14.3. Number of offspring produced by males versus females of given year-classes over their life spans if habitat were doubled or tripled

males occurs over a relatively short time period and the advantage of females, extending over a longer period, declines to that of males by the first reproductive period prior to doubling of habitat (fig. 14.3). The results are a corollary of the results of the effect of sex-ratio manipulation on yield (fig. 9.5). Doubling of the habitat is the exact equivalent of reducing the population in the existing habitat from zero recuitment to I carrying capacity, which maximizes recruitment. Female fawns do not breed at this population level, but they do begin to breed if the population is reduced further. The success of reproduction of females relative to males is strongly dependent on age of first reproduction. Note that the sexes of offspring in the empirical data occurred in equal proportions at approximately I carrying capacity, which would be consistent with these results (fig. 4.9). I carrying capacity is 99 and the 1:1 sex ratio of offspring occurred at 96–98, depending on whether the regression with a sample size of 18 or 17 was used.

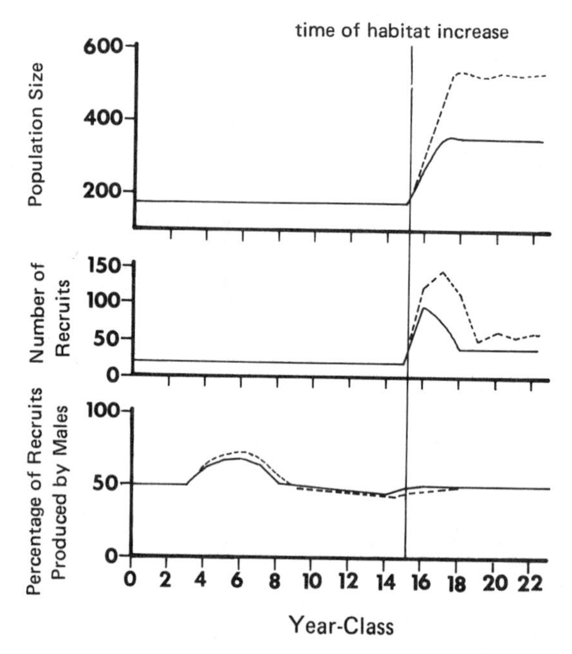

FIG. 14.4. Relationship of change in population size, number of offspring recruited, and relative values of males and females of a given year-class if habitat doubled (solid lines) or tripled (dashed lines)

A sudden tripling of the habitat is the functional equivalent of reducing the population below I carrying capacity. When this condition is simulated with the empirical population model (figs. 14.3 and 14.4), males of year-classes born prior to the tripling of habitat receive an even greater advantage over females than that obtained with doubling the habitat; but females have a disproportionate reproductive success following the tripling of habitat. The males of the influenced year-classes born prior to the tripling produce 842 recruits, while females produce 689. The equilibrium ratio would be 1.22 males:1 female (55 percent male). Year-classes of females born after the tripling of habitat and influenced by the change (i.e., before the value of males and females returns to equality) produce 351 recruits, while males of the same year-classes produce 279. Equilibrium ratio would be 0.79 males:1 female (44.3 percent male). Note that these disproportionate successes of the sexes occur even though each sex contributes 50 percent of the genes to each offspring.

Given periodic, sudden increases in K carrying capacity, males born prior to the increase will have disproportionate reproductive success, while females, depending upon the amount of new habitat, might have disproportionate success following the increase. Furthermore, the disproportionate advantage of males prior to increase in habitat and females following is directly related to the increase in K carrying capacity (i.e., the quantity and quality of new habitat). The greater the increase in K carrying capacity, the greater the disproportionate advantage of males before and females after the increase. However, because the males have a disproportionate success, the overall advantage is in favor of males.

In populations of deer at K carrying capacity in the absence of the fire or other events creating new habitat, the success of males and females would be equal. But if a fire or other habitat-creating factor occurred, males born in the year-classes prior to the fire have a disproportionate success. Therefore, given the periodic occurrence of creation of new habitat, the long-term average success of males in populations at K carrying capacity would be greater than females, and selection would favor those parents producing a disproportionate number of males. Also, the higher dispersal rate of males and the probability of locating patches of favorable habitat would further enhance the value of males at K carrying capacity. If a fire occurred and resulted in more than doubling of the K carrying capacity, females would have a disproportionate success following the fire, and selection should favor the production of a disproportionate number of female offspring. Indeed, female offspring should predominate under any circumstances (such as very high mortality induced by man, natural predators, and virulent diseases) that cause the population to fall below I carrying capacity.

These conclusions pertain only to the relative values of each sex before and shortly after the creation of new habitat. Complete assessment must consider a complete cycle of habitat, from the destruction of the climax vegetation through the successional stages leading back to a new climax. In the previous example, it was assumed that equilibrium occurred at the new and higher K carrying capacity. It is more realistic to assume that a gradual return to the initial K at climax vegetation would occur. Just as population growth due to increase in K carrying capacity gives disproportionate reproductive success to males because of difference in the time of life when reproduction is expressed between the sexes, declines in K carrying capacity would favor the disproportionate success of females. Assuming equal beginning and ending K carrying capacity, the net reproductive success of all males af-

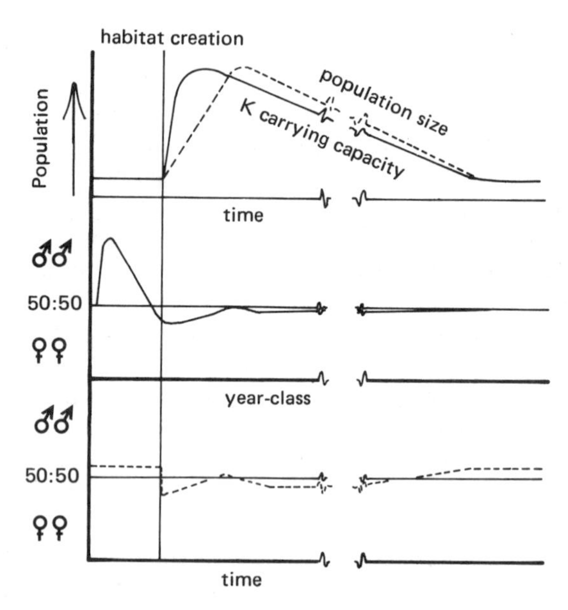

FIG. 14.5. Schematic relationship of a habitat cycle (from climax through succession back to climax) to the reproductive success of the sexes and the expected sex ratio based upon differential success

fected by the habitat cycle would equal the net reproductive success of all females affected. Therefore, a Fisher equilibrium would hold overall, and because males are the less expensive sex to rear (due to greater mortality during the period of parental care), sex ratio should show a disproportionate number of males, overall.

The overall shifts in reproductive success and expected sex ratios in relation to K carrying capacity are shown schematically in figure 14.5. Conclusions about shifts in sex ratio and reproductive success can be summarized thusly:

1. Because habitat creation has occurred frequently enough in the past to select deer for adaptation to subclimax conditions, and because males have a very disproportionate success when a disturbance occurs, natural selection should favor those parents giving a disproportionate amount of parental investment to males when the population is in balance with K. Therefore, the male sex ratio at conception in populations at K carrying capacity should exceed that predicted by the Fisher equilibrium.

2. Following the sudden creation of favorable habitat, if the K carrying capacity is doubled or greater, females immediately will

have a greater value than males, and selection should favor conception of a disproportionate number of females—a deviation in the opposite direction from that predicted by the Fisher equilibrium.

3. As K carrying capacity declines to the original value, females will have a disproportionate success, and natural selection should favor parents making greater investments in females than would be predicted by the Fisher equilibrium.

4. Over the entire habitat cycle, the net value of males should exactly equal the net value of females according to the prediction of the Fisher equilibrium. Because habitat is created suddenly and lost gradually, the advantage of males should be of shorter duration and greater amplitude, while that of females should be of lesser amplitude and greater duration.

5. Time lags in population response to shifting K carrying capacity will complicate the model, but not change the basic relationship.

6. All of the cited shifts in sex ratio can be encompassed in an adjustive, faculative mechanism that is cued by some factor relating population size to K carrying capacity. It should be noted that pronounced deviations in sex ratio of offspring could be expected (and more readily measured) close to the time of habitat creation, because with the sudden increase in K carrying capacity the relative values of the sexes varies greatly. The gradual nature of succession would result in more subtle sex-ratio differences that may defy statistical demonstration during return to climax.

This hypothesis may give a clue as to why young female deer produced more male offspring than older females, as reported by Robinette et al. (1955) and McDowell (1959). A male offspring is a long-shot gamble that pays off big if a fire or other event creates new habitat, or if the male disperses and finds a patch of favorable habitat. It may be that older females, by virtue of their larger size, better condition, and higher social standing, are better able to place stronger and better surviving female offspring into the highly competitive K carrying capacity home environment. They would also be expected to produce highly competitive males, but males are more quickly left to their own devices and must find their way in the world without further benefit from the mother. Young females in the K carrying capacity situation, unable to place female offspring in the competitive home environment, may be forced to gamble on the long-shot offspring. Therefore, in populations at or near K carrying capacity, older females would be expected to produce male and female offspring in more nearly balanced proportions, while young females would produce a disproportionate number of male offspring. However, if K carrying capacity were more than doubled or the population were reduced below I carrying capacity, nearly all offspring would survive; and since female offspring have a disproportionate success, both young and old females would be selected to produce a disproportionate number of female offspring.

Critical tests of this hypothesis must await further research. However, comparative evidence for other species can be taken from the literature. Although I have not made an exhaustive search, several papers have come to my attention that seem to support this interpretation. Mech (1957*b*) reported that wolf pups from a high density population in Minnesota had a disproportionate number of males, while an intermediate density population had an approximately balanced ratio; a low density population had a disproportion number of females. These results on wolves exactly parallel those for deer obtained from this study. For the vole (*Microtus ochrogaster*), Myers and Krebs (1971) concluded that certain genotypes producing progeny with sex ratios that deviated from 1:1 also had differential advantages depending upon population density. At low densities, selection favored the genotype that produced more females. Terman and Sassman (1967) reviewed the sex-ratio evidence for deer mouse populations and found that the sex ratio of *Peromyscus maniculatus* populations varied significantly with population density (with females being less prevalent at low density than at high density), while those for *P. leucopus* did not vary with density. *P. maniculatus* is typically associated with disturbed habitats (an animal "weed"), while *P. leucopus* occupies more stable, wooded habitats (Baker, 1968). L. L. Master (personal communication, 1978) has found that females of *P. maniculatus* predominate in the first, spring litters at times of low population density, while males predominate in the last fall litter at times of high population density. With a rapidly reproducing species such as *Peromyscus,* the advantage of the sexes can be expressed over seasons, while for long-lived, slowly reproducing species such as deer and wolves, it is expressed over years. A final significant result comes from those species in which sex, itself, is facultatively determined. The review by Anderson (1961) showed that most such species produced a disproportionate number of males at high population density. In conclusion, there seems to be a growing body of evidence that supports the hypothesis proposed here.

CHAPTER 15

Data for Management

Potential Indices to Population Status

Although the basic principles derived from this study are probably general to K-selected species, it is not known to what extent the population statistics relative to K carrying capacity obtained on the George Reserve may be typical of other white-tailed deer populations. For example, fawn females on the George Reserve begin to breed at about I carrying capacity, when the population is reduced from the K direction. If this is true of other populations, then the presence of breeding by fawn females would suggest the relative density of the population in relation to resources. A number of such parameters can be derived from the George Reserve deer population to serve as indices to population status elsewhere. They would seem to be useful as general approximations or broad guidelines. Whether they are reliable for more precise management decisions must be determined by comparative work. However, if agreement were reached by several or all of the indices, greater credence could be given to their use. For example, if breeding by fawn females, maximum age, and recruitment rate all suggest a similar relative density, then there would be good reason to believe that the relationships found on the George Reserve apply. One would expect to find the greatest compliance among populations with similar genetics and habitat characteristics.

Figure 15.1 shows the various parameters plotted against percentage of K carrying capacity on the x-axis. Parameters selected are those which are relatively easily obtained from hunter kill or road kill samples. Other parameters could be obtained from the data if they were appropriate to a given set of data available to the manager.

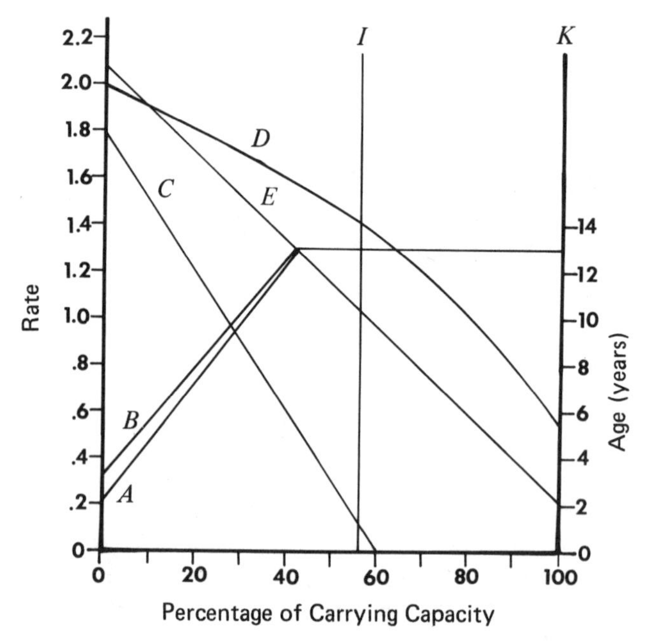

FIG. 15.1. Relationship of population parameters from the George Reserve deer population to percent of K carrying capacity. *A* is mean oldest males, *B* mean oldest females, *C* embryos per fawn female, *D* embryos per yearling and older female, and *E* recruitment rate.

Age Structure and Population Model

Little can be determined about the status of a population from age structure alone (Caughley, 1974; McCullough, 1978). Without some measure of population size or density, the interpretation of life-table parameters is virtually impossible. For example, a small rate of increase may indicate that some factor in the environment necessary for reproduction is no longer being met and the population is declining towards extinction, or it may mean that the population is near maximum density in a complete environment. Given only age-structure data, one could not distinguish between a population on the verge of extinction and a population at K carrying capacity. The concept of a typical life table for a species is a misleading one. The life table of a given population at a given time is valid only for that particular set of circumstances of environment and population characteristics. The number of different life tables which are possible under changing conditions is almost infinite; relative shapes of age-structure pyramids are indicative, but not diagnostic, for they allow too many alternative interpretations.

The first lesson for managers is apparent: one needs both age structure and population size. A valid index to population size or an estimate of density are reasonable substitutes for population size (McCullough, 1978), but given that age structure and population size are known, it is not possible to predict how the population will respond to changes in harvest. These predictions can only be made with some concept of how the population responds to exploitation—in short, a population model. Hence, the second lesson for the manager is that a population model is an integral part of any mangement program. In a real way, the purpose in determining age structure and population size is to develop a population model. Otherwise, the distribution of age-classes in a population is purely descriptive and has no utility. Most models used in present management are word or rule-of-thumb models. Improved predictability will depend upon quantification of population information.

Derivation of empirical models requires that the population be observed at more than one equilibrium state. Thus, it is necessary that different harvest intensities be conducted, and the manager should attempt to follow a rationale of management and collection of statistics that will lead to the formation of a model.

Determining Population Parameters

Although numerous attempts have been made to apply life-table methods to the analysis of kill data (e.g., Quick, 1960; Eberhardt, 1969), most of these methods have not proven to be useful at the practical level. One of the problems is the assumption of a stable age distribution. Many environmental factors operate on the population in unpredictable fashion, and competitive interactions within the population affect age-classes differentially. More fundamental, however, is the demonstration with George Reserve data that even when the variables were controlled and an S-shaped growth curve generated, a strict, stable age distribution was not achieved. Nor was it likely that one could be achieved under equilibrium conditions, except at populations near I carrying capacity. Thus, the first problem with the use of life-table analysis is that the assumptions underlying the method are seldom met.

A second difficulty is the lack of any determination of the actual population numbers from life-table parameters. Assuming that the life-table assumptions have been met, that the data are unbiased, and that the sample size is adequate, one still cannot determine the size of the population from which the sample was taken. It might have come from a population of one thousand, ten thousand, or one million. Theoretical ecologists have preferred the dynamic (horizontal) life table to the time-specific (vertical) table because of its clearer assumptions and mathematical consistency. The problems of the applied ecologists in reaching some sort of conclusion about the status of the population from the life table are the same with either table. The age structure in a given time-specific table may not be the same as that of a year earlier or later. Similarly, the survivorship of a given year-class throughout its life span in the dynamic table does not necessarily indicate survivorship of the previous or subsequent year-classes. In fact, a year-class that deviates greatly from the average in a limited environment is likely to be offset by a following year-class that deviates in the opposite direction because of compensatory processes. Theoretical ecologists appreciate that, in dynamic tables, the proportion of animals in a given age-class *never* exceeds those in younger age-classes; in time-specific tables, strong year-classes exceeding those of previous age-classes occur commonly. This raises havoc with the assumption of a stable age distribution. However, to the applied ecologist, the existence of a strong year-class is valuable information, and exploitation can be geared to take advantage of that year-class. The assumption of a stable age distribution for the elegance of a life-table calculation seems trivial by comparison.

Another major difficulty of the dynamic life table for the practitioner is the length of time required for its completion. One must usually make a decision each year about the exploitation to be allowed in the following year. Few wildlife populations have a span so short that a dynamic table could be completed in this time interval. Also, the effort required to obtain dynamic table data is typically great.

All of this leads to the conclusion that for regularly exploited wildlife populations, the time-specific table is more valuable to the manager than the dynamic table. Ironically, it is the least likely to meet the assumptions of the formal life-table calculations. Yet, for the purposes of the manager, the formal results are of relatively low interest. The primary concern is the extent to which the results are representative of the actual population, the comparison of the current rate of exploitation to MSY or FRY, and the assessment of the population relative to I and K carrying capacity.

To be useful in assessment of the living population, data from the kill (or animals found dead) must represent a reasonable sample size. If the kill is trivial, the sample will be inadequate and reliability will be low. The use of hunting to maintain a population near I carrying capacity results in the highest sample size and the greatest reliability for the analysis of kill statistics. If a trivial kill is taken from a large population near K carrying capacity, the sample is not likely to be representative. And, since irregular fluctuations and/or catastrophic events are common in populations near K carrying capacity, the predictability of subsequent population response is low. The predictability of kill data is greatest near I carrying capacity and lowest near K carrying capacity, just as with population models.

If vulnerability of animals by age and hunter selectivity are completely absent, the sample obtained in the kill will be representative of the living population, and the sum of individuals in each age-class can be considered to be a l_x series. Since these conditions are almost never met, a better estimate of l_x is derived from adding the d_x values of the oldest age-class to the age-class for which l_x is desired. Then, an estimate of the l_x series is obtained which is corrected for selectivity.

If hunting is the sole cause of mortality, the l_x series can be used to derive an estimate of the pre- and posthunt population. This has been done in table 15.1 for the George Reserve deer data. Since the data for individual years involve small numbers and unrepresentative samples, the accumulated results for nineteen years are used as if they were taken in one year. It can be seen that the results are reasonable approximations of the actual population data and population parameters. The method assumes that the population is momentarily stationary (i.e., no change in population size) but not that there is a stable or stationary age distribution between years. The critical assumption is that hunting is accounting for all mortality and the sample of the kill is representative. Note that error in the first assumption always leads to an underestimation of the actual population, while errors in the second lead to biases in either direction. Overrepresentation of young animals leads to underestimation of actual population, while overrepresentation of old animals leads to overestimation. Rarely does a single factor, such as legal hunting, account for all of the mortality. Under management for MSY, however, it can greatly predominate, as has been demonstrated on the George Reserve. Usually, a number of factors other than legal hunting contribute to total mortality in a deer

TABLE 15.1. Comparison of Actual (Observed) and Predicted Populations for the Combined Nineteen-Year Kill on the George Reserve (♂ and ♀ Kill Combined)

Age Class	*Kill*	*Prehunt Actual*	*Prehunt Predicted*	*% Error*	*Posthunt Actual*	*Posthunt Predicted*	*% Error*
0	246	822	832	+1.20	576	586	+1.71
1	298	580	586	+1.02	282	288	+2.08
2	98	289	288	0.00	191	190	−0.52
3	94	191	190	−0.01	97	96	−1.03
4	51	97	96	−0.01	46	45	−2.13
5	13	45	45	0.00	32	32	0.00
6	8	34	32	−5.88	26	24	−7.69
7	13	25	24	−4.00	12	11	−8.33
8	5	11	11	0.00	6	6	0.00
9	2	7	6	−14.29	5	4	−20.00
10	2	3	4	+25.00	2	2	0.00
11	0	3	2	−33.00	2	2	0.00
12	2	2	2	0.00	0	0	0.00
	832	2109	2118	+0.42	1277	1286	+0.70

Actual recruitment rate = 0.6437 Kill/prehunt N = 0.3928
Estimated recruitment rate = 0.6470 No. ff/prehunt N = 0.3928

population. To deal with this difficulty it is necessary to introduce the concept of a "pool." The pool is defined as some given segment of the population from which the harvest is taken. It is defined in time and space and by specific characteristics. The pool from which the George Reserve deer kill is taken is those animals, within the fence, of approximately six months of age and older that were living between October through February when the kill is made. At I carrying capacity or less, the pool is composed of the entire population; but at higher densities, the pool constitutes a smaller fraction since the mortality of young early in life and chronic mortality remove animals from the pool (which is subjected to hunting). In areas where bucks-only hunting is allowed, the pool would be defined to exclude females; however, little could be determined about the population from age structure of the kill since the females that produce the recruitment are excluded (tables 9.2 and 9.3).

The value of the concept lies in the fact that the age data obtained from kill will represent the pool even though it may be entirely unrepresentative of the total population. But since the pool, as defined, is the object of interest to the manager, this is the most relevant information. Note that the George Reserve kill data represents only animals reaching recruitment age. Estimation of total numbers of offspring born are derived from embryo counts, not from age at time of death. The pool can be modeled for exploitation on the basis of data derived from the kill. Although it is highly desirable that other information (for example, an independent estimate of total population size) be obtained, such information can be treated as a "black box" that behaves in some functional relationship to the pool. Thus, if the hunting kill is very small, a small pool will be indicated (i.e., most mortality will be accounted for by factors other than hunting). Liberalizing the kill will increase the take and increase the size of the pool. Because of compensatory processes, the increase in the legal kill will likely come at the expense of some other mortality factors. If the kill is approximately stabilized at a given number and held to that number for a period of time, the mean recruitment to the pool will tend to come to equilibrium with the kill (see chap. 9). Since the size of the kill is known, the total size of the pool can be calculated by dividing the mean kill by the percentage of recruitment as derived from the kill data. If all sex- and age-classes are legal, and if the kill is relatively high, the pool will approximate the total population. The importance of being able to regulate the kill within narrow limits should be obvious. If males only are killed, little about the total population can be derived from analysis of the kill.

When changes in the size of the kill are made, they should be made in relatively small increments to minimize time-lag effects. If larger increments of change are introduced, the shift in the recruitment rates derived from successive kills should be directional, but gradually return to equilibrium as long as MSY is not exceeded. Small increments of change are clearly preferable. Similarly, the new level should be maintained long enough for equilibrium, on average, to be clearly established. Thus, year-to-year variation in recruitment must be expected due to unidentified variables, and enough years must be accumulated to obtain confidence in the estimate of the mean. Again, the least stability, and hence greatest variance, can be expected near K carrying capacity and the most stability near I carrying capacity. If the population is approaching I carrying capacity, one must be sure that an equilibrium has been achieved before an additional increment in kill is attempted. This strategy of management could be applied to any K-selected species, as long as compensatory processes are present in the

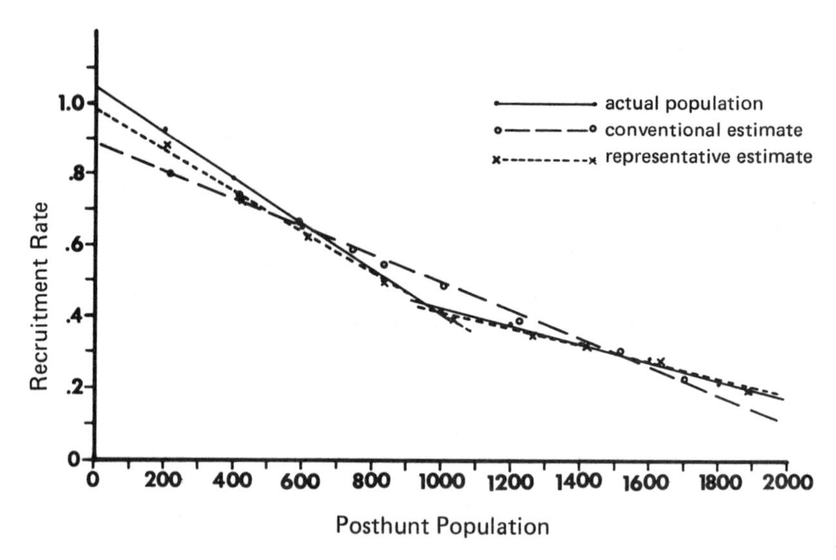

FIG. 15.2. Comparison of actual recruitment rate on posthunt population size of an artificial population of seed beads with estimated recruitment rate from conventional, life-table analysis and representative methods

population. Indeed, K-selected species, almost by definition, are those which adjust to resources in this manner.

Such an approach to age-structure analysis was further tested by creating artificial populations of seed beads. The reason for using seed beads instead of the computer was to mimic the case in nature where certain age-classes are vulnerable and others are not, and this information was not known in advance by the researchers. The seed beads allowed large numbers to be encompassed in a small container, and each age-class in the population could be represented by a different color of bead. Because the beads varied by color as to size and shape, it was not known, a priori, how they might segregate during mixing. Beads of some colors were slightly larger than others, and some were round while others were oblong or tubular. The population was created on the basis of a two-segment linear regression, as shown in figure 15.2, and age-classes were assigned according to their expected frequency based upon the George Reserve data. The model was purposefully made complex in order to complicate the problem and give a rigorous test of the system of analysis. Popuations were created at 200-bead intervals, from 200 to 1800, and placed in a large tin can. A plastic snap-on cover closed the can and it was shaken and rolled in all

TABLE 15.2. Method of Calculating Population Parameters on One Sample of Kill on a Seed Bead Population of 1800

		Conventional Life-table Method[1]	*Representative Method*[2]
x	*Kill Number*	*Population*	*Population*
0	52	313	313
1	42	261	253
2	26	219	157
3	30	193	181
4	26	163	157
5	29	137	175
6	25	108	150
7	27	83	163
8	12	56	72
9	18	44	108
10	12	26	72
11	9	14	54
12	5	5	30
13	0	0	0
TOTAL	313	1622	1885
Estimate of:			
a. Recruitment Number		313	313
b. Population producing recruits		1309	1572
c. Recruitment rate		.2391	.1991

[1] Kill numbers summed from the bottom of the column to derive l_x series.

[2] Age-class proportional to that in the kill, with the zero age-class being the standard. Thus, for age-class three, 52:313 as 30:x, x = 181.

directions in an attempt to completely mix the beads. Samples (the harvest) were selected by blindly dipping a spoon into the beads and spreading the beads out on the table to be separated into colors. The sample harvest was made to equal the recruitment, i.e., the number of zero-age class beads in the population. This was done on the assumption that if the kill were fixed over a series of years, an average recruitment of the pool would come to balance the kill. Once the sample was separated and counted, it was returned to the population and another sample was drawn by the same procedures, until five samples were obtained (mimicking five years of fixed kill). The samples were then combined into one estimate of the population size and recruitment rate. Methods of calculation are shown in table 15.2. The conventional life-table approach, of course, assumes the existence of a stable age distribu-

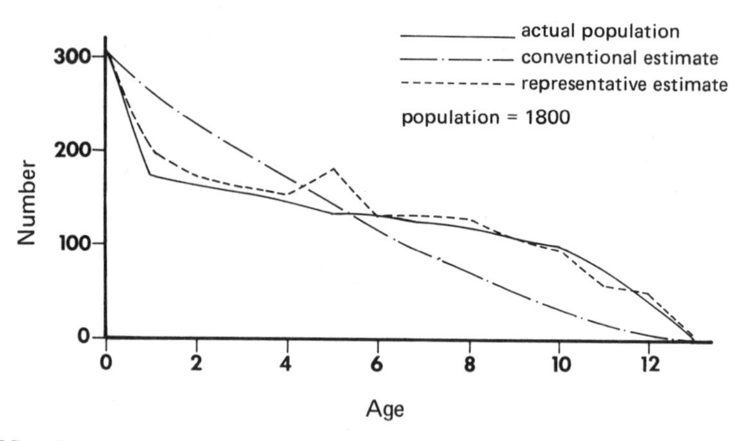

FIG. 15.3. Relationship of actual "age structure" of an artificial population of seed beads compared with that estimated by conventional life table methods and representative methods

tion, while the "representative" age structure does not. Note that the sample in table 15.2 shows nonrandom selections (fig. 15.3). Thus, the desired bias of "vulnerability" of certain age-classes occurred in the exercise.

Figure 15.3 illustrates where error comes in as a result of not meeting the assumption of a stable age distribution. The actual age distribution at this population level—near subsistence—is not stable, and following conventional life-table methods results in a distribution of age-classes that does not match the actual. The representative approach, however, gives a reasonable approximation of the actual distribution, the errors being due to nonrandom selection (vulnerability) of certain colors of beads (age-classes). The resultant population parameters calculated from the sample show similar relationships (table 15.2). Actual population in table 15.2 was 1800, while the conventional life table gave an estimate of 1622 and the representative table 1885. Recruitment-rate estimates were 0.2391 and 0.1991, as compared to the actual value of 0.1739. Once again, the representative approach gave better estimations than the conventional life-table method. Since the objective is to derive a recruitment model, it is important to note that errors in population size and recruitment rate are compensatory. Thus, estimates of population size which are too high give an underestimate of the recruitment rate, while estimates of population that are too low give estimates of recruitment rate which are too high. Hence, a

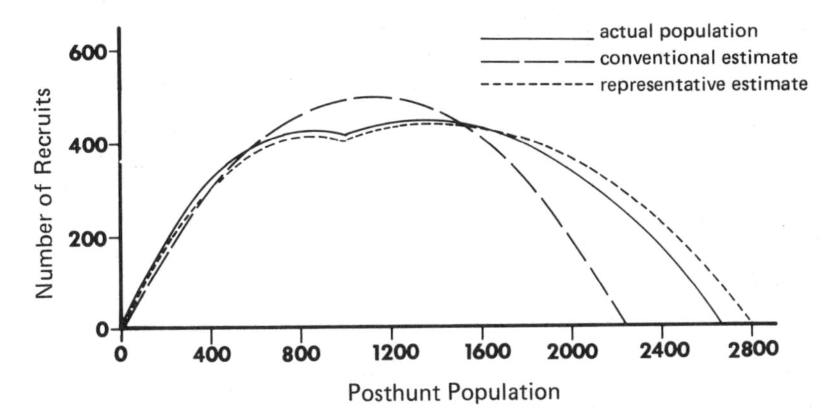

FIG. 15.4. Parabola of numbers of recruits on posthunt population size of an artificial population of seed beads as derived from the regressions in figure 15.2

recruitment model derived from these data will be better than if the errors were independent.

The outcome of these seed-bead populations is shown in figures 15.2 and 15.4. Note that the conventional estimate gives a single recruitment-rate regression, and this would be true if the actual population model had been a single regression. As mentioned in the age-structure section of chapter 4, the stable age distribution will be approximated at the mid-values, near MSY. If the population model had consisted only of the regression on the left-hand side of figure 15.2, the MSY would have occurred at about 810 and the population would have had approximately a stable age distribution at a population of 625. However, if the population model had consisted of only the regression on the right-hand side of the figure, the MSY would have been about 1340 while the stable age distribution would have been approximated at about 1520. Thus, as b increases (smaller negative number), the expected point in population size where a stable age distribution is approximated shifts upward. The representative method gave estimates of both population size and recruitment rate that approximated the actual values, and the models derived gave reasonably good predictions.

The rationale, value, and problems of interpretation of this approach to age-structure analysis can be summed up as follows:

1. The critical assumption is that the number recruited comes to balance with the number killed. Or, if a number of other mortality factors are operating in the system, that these are reasonably

constant and bear a consistent relationship to the hunting kill. This will be true if the regression of recruitment rate on posthunt population size shows a good fit. The number of years at which a fixed kill must be maintained in order to assume that the mean recruitment equals the mean kill depends upon the variance and increases as variance of recruitment rate increases. In an unknown case, variance of recruitment rate can be determined from age-structure analysis between years at a given kill. The kill is an external variable that can be controlled to introduce stability, and most agencies have the legal prerogative to do so. However, social and political constraints may prevent a precise approximation of this ideal, and the greater the variance in kill, the longer the necessary series of years before the mean kill will have acceptable confidence limits. Mean recruitment rate and mean population size can be derived from the composited sample.

2. Distribution of the age structure of the kill can be plotted as in figure 15.3 and compared with the conventional estimate based upon the assumption of a stable age distribution. If they correspond, then either method will give a good estimate. If the sample differs from the conventional estimate, the representative estimate should be used. Note that population parameters can be derived by the representative method with only two age categories, the zero age-class and the combined, older age-classes. Thus, if it can be assumed, a priori, that a representative kill is obtained, age determination of the kill sample is simplified. If the assumption cannot be made, ages of all classes are necessary to compare the two methods.

3. If the plot of the age structure is not representative (which in most cases would involve too few animals in the zero or one age-classes because of discrimination of hunters against young animals or vulnerability, as with the George Reserve kill data, one would follow one of two routes. If the conventional estimate approximated the older age-classes, the conventional estimate, which also estimates the younger age-classes, could be used. If the conventional estimates did not approximate the older age-classes, then the older age-classes would have to be treated by the representative method and be considered as a pool modeled separately from the total population, with the first age-class being treated as the recruitment age-class.

Clearly, there are many assumptions and problems in using age-structure analysis to estimate population parameters. In many cases so many unidentified variables are operating that the results would be incomprehensible. But the methods of population census currently in use are, by and large, so poor and the data necessary for this analysis so readily obtained that it would be pointless not to examine the possibility. And in situations where MSY or FRY was the objective of the program, hunting kill would be a predominant mortality factor, age structure would be approximately representative, and this approach should begin to give interpretable results. As such results are incorporated into a model of the population, further predictions would be generated. Assuming management proceeds carefully and incrementally, the feedback system of data set, model and hunting program, new data set, etc., should lead to a better achievement of objectives.

While the approach can be demonstrated successfully on artificial populations, a full-fledged field demonstration has yet to be attempted. It is not possible to conduct this test on the George Reserve because the small kill per year (therefore, small sample size) results in poor estimates of the distribution of age-classes in the kill. Obviously, proof of usefulness of the method in actual practice must come from a larger land unit. Furthermore, from the management point of view, the method is fraught with difficulties, like adjusting the hunting regulations to achieve a fixed kill of desired size. Also, the assumption that habitat is approximately constant (either it is stable or being created at the same rate as it is being lost) may be difficult or impossible to meet in practice. Recognition of the importance of these factors, however, is the first step. Current procedures used to reach management decisions are rather arbitrary and usually lacking in verification. Even if the approach suggested here should ultimately prove to be impractical in practice, the rationale will give a better understanding of the general relationships involved and suggest the shifts in population parameters that should accompany changes in management.

CHAPTER 16

Management Concepts

Gross (1972) has pointed out that intuition can lead to erroneous conclusions about big-game management, and that intuitive criteria have often been used in making management decisions. The empirical results of this study underscore the fact that many of the responses of deer populations are counterintuitive. The points Gross makes concern reproductive rate, population density, nutritional density, and range condition. Intuition has often lead to the belief that reproductive rates should be maximized through management, but to do so with the George Reserve deer herd would lead to near extinction, since reproductive rate increases linearly with reduced density. Thus, assuming MSY is the intended objective of the program, what the manager should attempt to achieve is the best trade-off of reproductive rate and population density. Intuitively, high population densities are desired, but the highest densities have the lowest rate of productivity. Intuitively, competition should be eliminated, but this is incompatible with maximum density. Complete elimination of competition occurs only at very low levels, as demonstrated by linearly increasing reproductive rates on population density to the y intercept. Intuitively, range condition should be ideal; in practice, this has typically led to the belief that the minimum impact on the vegetation is desirable. In fact, this state can only be achieved with no grazing, since even light grazing will have an impact on the plants most susceptible to grazing and such species will decrease, while plants less susceptible to grazing will increase. Thus, maximization of yield will, inevitably, lead to some state of the vegetation which is different than it would be in the total absence of grazing.

Gross (1972) attributes the counterintuitive behavior of populations to their characteristics of nonlinearity, thresholds, and limits. These, I believe, are symptoms and not causes, since the existence of such characteristics are commonly recognized, and have been for a long time. Given that managers have long known of these characteristics, why have they failed to note the inconsistencies in their criteria? Obviously, one difficulty is the failure to comprehend the dynamic interaction between various population parameters as density changes through a broad range, and this failure is due, to considerable extent, to the lack of empirical evidence. Studies thus far have been too piecemeal for the complete interactions to be demonstrated. The response of many readers to the outcome of the George Reserve deer studies will be that they were obvious and expected, but as with most Monday morning quarterbacking, the obvious was far more apparent after the fact than before. My own uncertainties at the outset and those I have observed in discussions with astute and experienced biologists have convinced me that the outcome was hypothetical at best, and certainly not obvious.

A second, and probably more important explanation for the failure of intuition was the failure to clearly distinguish the difference between I and K carrying capacity. Reexamination of the criteria of management previously cited will show that for reproductive rate, lack of nutritional competition, and range conditions the emphasis is on I carrying capacity, while for population density, emphasis is on K carrying capacity.

Harvestable Surplus versus Sustained Yield

A fundamental underpinning of the wildlife management field, ever since its cogent formulation by Aldo Leopold (1933), is the concept of the surplus or harvestable crop. The idea is based upon the simple observation that most animals produce more individuals than are necessary to maintain the population, and this extra number can be removed by hunting without detriment to the population. That this idea is still current is demonstrated by the fact that Leopold's *Game Management* was commonly employed as a text in wildlife management courses up until a few years ago, and "the surplus" is used as a major justification of the biological soundness of hunting to antihunting groups.

While the concept of a surplus is correct, unfortunately it is a static one that fails to encompass the dynamic and compensatory nature of population responses which are the roots of the potential for exploitation. Because recruitment tends to come to balance with the kill, the idea of the surplus tends to lead to the conclusion that what has been done in the past is the correct thing to do in the future. Consider the case of a deer population which is near K carrying capacity. At this level, few young survive to recruitment age. Since the recruitment is small, the indicated surplus that could be taken would be small. Management according to the surplus criteria would result in light exploitation, allowing the population to continue at this low rate of productivity but with a high standing crop. Using survivorship to yearling age, a common criterion in big-game management, would result in a similar outcome. On the other hand, suppose that a population were reduced by a catastrophe, or that a major improvement of habitat occurred due to logging or fire. Observation of the high survivorship of young to recruitment age under such circumstances would suggest a large surplus, and a large harvest could be taken. Again, the current state of the population would likely be maintained, so that future management would tend to be dictated by the effects of prior management. It should be obvious that surplus is as much (indeed more so) a characteristic of the management program as it is of the population.

In actual practice, many wildlife managers recognize the potential of populations to sustain kills above that of the surplus and manage accordingly. Yet the concept of surplus persists, as is well illustrated by Caughley's (1974) comment on a paper in which the authors used a "surplus" criterion for determining the allowable exploitation rate. In spite of Caughley's pointing out the relationship of population size to MSY, one of the authors of the original article, in response, restated the conclusion based on the concept of surplus (Schladweiler, 1974). The example is brought out not as a specific criticism of the worker, since a good many wildlife managers would have responded similarly, but to demonstrate that the concept of surplus has, indeed, become a real problem. Leopold's articulation of the concept was a giant step forward at the time, and it served the field well for many years. But our knowledge has increased since then, and, unfortunatley, adherence to the concept of surplus has become an obstacle to further advancement of the art.

Similar difficulties apply to fixed rules of thumb about management. For example, it is common for deer-management biologists to use 25 percent of the prehunt population as the sustainable kill even though it is just one, arbitrary point of a continuous variable and ignores the relationship between the percentage of the population removed and absolute numbers relative to MSY. Table 16.1 shows the relationship of population size to the percentage of kill

TABLE 16.1. Relationship of Population Size to Percentage of the Population Represented by the Kill at Equilibrium for the George Reserve*

Posthunt Population	*Prehunt Population*	*Number Killed*	*Kill as a % of Prehunt Population*
10	19.37	9.37	48.37
20	37.74	17.47	47.01
30	55.10	25.10	45.55
40	71.47	31.47	44.03
50	86.84	36.84	42.42
60	101.21	41.21	40.72
70	114.58	44.58	38.91
80	126.94	46.94	36.98
90	138.31	48.31	34.93
99	147.69	48.69	32.97
100	148.68	48.68	32.74
110	158.05	47.53	30.07
120	166.42	45.45	27.31
130	173.78	41.75	24.02
140	180.15	36.30	20.15
150	185.52	29.99	16.17
160	189.89	19.68	10.36
170	193.26	8.24	4.26
176	194.80	0.00	0.00

*The boxed data indicate MSY.

for the George Reserve deer population. A kill of 25 percent of the prehunt population is one sustainable yield, but so is any other kill as a percentage of the prehunt population below about 48 percent. MSY for the George Reserve is achieved at about 33 percent kill, but it may differ in different areas—areas that will have a different set of mortality factors and different compensatory relationships between hunting kill and other mortality. A kill of 25 percent of the herd may be too great. The problem is that 25 percent is arbitrary. Perhaps the point can be brought home by a simple example of application of the traditional wisdom to the George Reserve. Suppose one had a measure of the prehunt population size and killed 25 percent of that number. If this kill was repeated over a period of years, there would be little change in population size over time. The traditional conclusion would be that 25 percent was a sustainable harvest. Now suppose the manager, out of curiosity, decided to begin killing 10 percent of the prehunt population. In the following years the population would increase. The traditional wisdom would lead to the conclusion that this kill was less than the population could sustain. Then suppose the manager decided to kill 30 percent of the prehunt population. The population would start to decline in the following years. The traditional wisdom would lead to the conclusion that the population could not support such a high kill. Therefore, the manager who followed traditional thinking would return to a kill of 25 percent of the prehunt population, secure in the knowledge that he has actually verified the correctness of that figure as the MSY!

These are just a few of many difficulties that exist because of the failure of the field of wildlife management to develop a rigorous body of theory based upon a tightly reasoned set of concepts. The profession cannot hope to contend with the myriad of challenges that confront it by clinging to outdated and easily undermined concepts. Unfortunatley, the early work of Scott (1954) appears to have had little impact. Encouraging signs that changes may be coming are the works of Gross (1969, 1972), Wagner (1969) and Caughley (1976). Hopefully the present work, by showing the relevance of theory to empirical results and vice versa, will give impetus to the movement.

Maximum Sustained Yield and Harvest Regulations

In chapter 9 the importance of sex and age ratios and intervals between harvests to MSY were considered. Designations of legal game (e.g., bucks-only, forked-horn or better) are tools by which management manipulates sex and age ratio. The reasons for such designations are as often socio-political constraints (hunter resistance to shooting females and young) as they are scientific. Yet, the concept of increased yield by unbalancing the population in favor of females is almost universally accepted on intuitive grounds. The result of imposing environmental constraints on the population is that the apparent advantages are not realized. In terms of MSY there is little (if any) advantage to be gained by manipulating sex and age ratios in the population. Obtaining any gains requires a fine degree of tuning of the estimation of population parameters and control over the size and composition of the kill. Such precise data are almost never available in wild populations. Control of the size and composition of the kill in sport hunting cannot transcend the ability and willingness of hunters to conform. Even well-meaning hunters will err, and such errors will always be part of the management system. Nevertheless, achieving MSY is not dependent upon precise adjustment of sex and age ratios. The important variable is the size of the kill.

For practical purposes, MSY can be obtained with unselective hunting. Thus, in terms of yield of numbers, no legal stipulation of sex or age need be specified in the regulations, for MSY is quite insensitive to sex and age composition of the population. Again, this outcome was counterintuitive. Note, also, that achieving advantages in yield by age-ratio manipulation would require selectively shooting a very disproportionate number of female fawns. Because sex of fawns is nearly impossible to determine in the field, a disproportionate kill of female fawns would require a similar high kill of male fawns. Although protection of older males would not be required, considerable protection of older females would have to be achieved. About three female fawns would have to be killed for each older female killed. The regulations to accomplish this end would be intricate, and the response of hunters to them would make an interesting study in social psychology. Perhaps such a program could be accomplished in a small-scale experiment, but it would seem impossible on the broader scale on which agencies typically manage.

Management for MSY with unselective harvesting also requires a fair kill of younger animals and, of course, many antlerless animals. But management of MSY implicitly requires that hunters are willing to kill such animals. It would seem to be more palatable to the hunters, in that each individual hunter would be allowed to take the best animal he could kill. Hunteres with special ability or access to privileged areas would be more likely to take an antlered male. All hunters could be selective for sex, males usually preferred. Many hunters would be willing to take the first (or only) deer encountered, or would take such animals as the season progressed and they were unable to get a larger male. Under any sort of MSY management, individual growth rates of deer would be high, and even fawns would be good-sized animals. Thus, hunters would be spared the embarrassment of the "jackrabbit" fawns that are frequent when an antlerless take is allowed on a population previously managed by bucks-only hunting. Finally, there appears to be no gain in MSY to be obtained by periodic harvests; however, there may be other benefits. Also, greater average yields could be obtained by periodic harvests from populations which had to be held to very low densities (i.e., below I carrying capacity) for other reasons, such as to reduce crop damage or deer-car collisions.

The simplest scheme is to manage for fixed-removal sustained yield (FRY, chap. 9). It requires giving up a certain amount of yield, but the advantage is that only a fixed number of animals need to be harvested each year, and the chance of overharvest is reduced. Less in the way of population information is required, and regulations can be essentially the same year after year, as long as the basic carrying capacity of the management unit does not change. Simplified regulations are more easily administered by agencies and are appreciated by sport hunters. No doubt compliance improves, for unintentional violations probably increase with complexity of regulations and amount of change from year to year.

Managing for Trophy Males

Trophy males, as used here, refers to large-antlered males, the head of which hunters would be inclined to retain for display purposes. In general, eight-point (total points) antlers with a heavy beam would be included in the category. It does not refer to record-book males, which are exceptions that occur very infrequently. These record-book males are so uncommon, and the number of hunters who take them so few, that they do not seem to be a reasonable goal upon which to base a management program. Be that as it may, it should be observed that the program that maximizes the number of trophy males has, over the long run, the greatest probability of producing an exceptional record-book-class male.

In the wildlife management field, trophy-male management is usually understood to be management by bucks-only hunting, and the resulting populations are near K carrying capacity. To what extent this is a euphemism applied to populations where hunter acceptance of MSY management could not be obtained is not clear. Whatever the case, the common association of trophy-male management with bucks-only, lightly harvested populations near K carrying capacity has obscured the biological basis of trophy male production. Production of numbers of trophy males is associated with trade-offs between high growth rates and old-age structure of males. The first condition is not met in bucks-only harvested populations near K carrying capacity, and that fact should be clearly understood. Growth rates are slow, and although age structure is old, the net outcome is poor. It takes twice as long to produce the same trophy at K carrying capacity as it does at I carrying capacity. Furthermore, the production of numbers of trophy males under bucks-only hunting goes way down. In chap. 9, in the section on manipulations of sex ratio, it was shown that an equilibrium yield of 6.79 males (best case) would be obtained by bucks-only hunting on the George Reserve (table 9.3). This would be obtained from a prehunt population of 179.67, so the kill would represent 3.78 percent of the population. Of course the 6.79 yield could be from the largest and oldest males of the population, with trophy heads from ten- to twelve-year-old males.

However, the route to producing large numbers of trophy males is to manage females for MSY and restrict the kill of males to trophy heads. This way, many more males are recruited and individual growth rates are relatively high. Even though the turnover of males is much greater than with bucks-only hunting, the harvest of males can be delayed until they are old. The only limitation is the ability of the agency to effectively protect males until they are old by specifying regulations on legal bucks. Thus, not only is the number of trophy males greatly increased, but the older age in conjunction with higher individual growth rates could yield equivalent trophy males, and the exceptional individuals would be superior.

For example, assume the George Reserve females are being managed for MSY. This would involve an initial posthunt population of 99 in which 51.75 were female (table 8-1). Females would recruit 26.00 male fawns and 22.69 female fawns. Stabilizing the females to a 51.75 posthunt equilibrium would require a kill of 3.46 female fawns and 19.23 older females. Since most hunters could not distinguish male fawns and yearlings from females in the field, a certain kill of these males would be a necessary cost of MSY management of females. The cost can be estimated from the ratio of male to female fawns killed on the reserve (1.32 males to each female) and the same ratio of yearling males to yearling and older females (0.47 males to each female). Thus, 4.57 male fawns and 9.04 yearling males would be removed from a recruitment of 26.00 male fawns. At equilibrium, 12.39 males would live through the yearling age, and assuming they could be protected by regulations between the spike-antlered and eight-point stages, this could be the trophy-male yield. It is almost double that obtained by bucks-only harvesting.

Because the weight and antler-size analyses have not been completed, it cannot be said at this time what the trade-offs between age and growth would be. Also, the average life expectancy would determine the size of the male population at equilibrium.

Although recruitment was independent of numbers of males in the posthunt population (chaps. 6, 9, and 14), at some higher density this relationship might change as increasing numbers of mid-aged males were retained in the population to produce old, trophy males. Furthermore, retaining males until older ages would increase the male population, and growth rates of males would decline due to greater competition. Thus, increasing the number of old, trophy males would decrease the average size of the trophies. The best combination of age, growth rate, and trophy yield probably would be obtained by harvest shortly after reaching trophy size. In any event, the trophy production by MSY management of femles would equal or exceed the trophy yield of bucks-only harvest with the additional yield of antlerless animals. In fact, the MSY of the population would be obtained, since the only change would be the age at which males were killed. The number of males killed per year (26.00) would not be different at the new equilibrium, although the total posthunt population size would be greater and contain a higher proportion of males.

Game Cropping

This book emphasizes management of sport hunting because the white-tailed deer is the most important big-game animal in North America, where sport hunting is the paramount use of big game. But the principles derived from the George Reserve deer studies have relevance to game-cropping schemes, since the basic questions are the same. How is MSY achieved, and what is the outcome of any given SY?

Caughley (1976) has reviewed game-cropping attempts conducted mainly in Africa. The concept behind game cropping was that the mix of natural herbivores had evolved with the vegetation and, therefore, should be a more efficient converter of plant biomass to animal biomass than exotic domestic stock (Dasmann, 1964*b*). However, Caughley concludes that most ventures in game cropping have failed, and a major source of failure was the overestimation of yield based upon populations at K carrying capacity. Other problems were nonbiological, such as problems with efficient killing, carcass preparation, refrigeration, and transportation. (Only the biological constraints will be considered here.) Furthermore, it was commonly stated by advocates of game cropping that game did not degrade the vegetation like domestic stock did. At K carrying capacity this is probably true, since inputs by man would maintain high populations of domestic stock over periods when wild species would be suffering high mortality. Such man-induced time lag would continue the pressure on the range, with an adverse outcome. Nevertheless, a comparison of wild ungulates with domestic ones at K is not relevant to their relative merits as producers of meat. Both will give MSY at I carrying capacity where range deterioration is unlikely in either case.

Caughley (1976) questioned the assumption of biological superiority of wild herbivores over domestic sheep or cattle as converters of plant tissues to animal tissues. He noted that domesticated eland (*Taurotragus oryx*) are less efficient converters than domestic cattle. This is understandable, since the eland is relatively few generations removed from the wild stock and still carries many genes adapted to meet the exigencies of living wild. Cattle have been selected for perhaps ten thousand years for growth and production, and man has assumed the role of equalizer of environmental limitations. If there are predators, man goes with the cattle to protect them, builds predator-proof structures, or kills the predators directly. The wild herbivore must watch, run, and hide from predators. If diseases and parasites are a problem, man employs sanitation, vaccination, dipping, and other practices. The wild herbivore must pay the physiological cost of immune responses, whether they are needed or not. If there is a bad season, man moves the cattle or feeds them supplements. Wild herbivores must fatten or migrate. Water deprivation and numerous other examples could be discussed. In essence, the process of domestication is one of selection for productivity, with man assuming the burden of protection. Protection in the wild herbivore involves a high physiological cost, and that cost comes at the expense of reduced production. It would be surprising if wild herbivores in early states of domestication were as successful at production as domestic stock. It would be equally surprising if domestic stock were as successful as wild herbivores if turned loose in the wild to face environmental extremes and a full complement of competitors and enemies.

Other factors influence the efficiency of conversion of wild and domestic herbivores. In taking yield from domestic animals, man can take animals according to age and reproductive value. He markets young at the end of their rapid growth period and old females, saving only enough young females for replacement. He incurs little chronic mortality in his herds since the age of individuals is approximately known. One wonders if some of the striking color patterns of domestic cattle were not selected by man because they allowed ready identification of individuals in the herd, therby allowing culling by age. Age manipulation in wild herbivores is difficult because the age of most of the population cannot be determined. The tendency is to harvest prime animals, not young and old. This, plus the confusion of mistaking young males for females, leads to constraints in manipulation of wild herbivore populations for maximum production.

Social behavior carries an energetic cost. It is probably not accidental that the successful domestic animals were derived from wild ancestors that formed aggregations with relatively little intolerance of crowding. This behavioral tendency was probably increased by selection for tameness in domestication. Solitary or secretive wild herbivores are poor candidates for domestication. Furthermore, most domestic and wild ungulates are polygynous, and the greatest behavioral costs pertains to sexual competition among males. Man controls the cost of male behavior in domestic stock by keeping males separated, except for the breeding period, and then introducing only enough males to breed the females. He then turns male offspring into efficient converters with femalelike behavior by castration. He need keep only enough unaltered males to replace his breeding stock, which, incidently, he markets when old. Thus, he manipulates sex ratio in the population and manipulates behavior of the male offspring. Wild herbivores must pay the cost of having many males in the population, and if niche separation between the sexes is present, it sets limits on what can be accomplished by sex-ratio manipulation.

Biologists have long recognized that large herbivores that feed on herbaceous parts of plants have a large but low average quality of biomass available to them. Long feeding times, relatively large energy expenditures in gathering and masticating food, and slow rates of digestion are required. By contrast, herbivores feeding on fruits and seeds and carnivores that eat meat spend much of their time searching. Once food is found or captured, however, it is of high quality and can be consumed and digested rapidly. McNab (1963) has referred to these two broad classes as "croppers" and "hunters," respectively. But these are relative terms, and the croppers show considerable variation that can be further categorized. There are what I call "swathers," with a broad tolerance for plant species and plant parts, and "gleaners," which select plant parts and sometimes species carefully. The American bison (*Bison bison*) is an example of the swather—it mows grasses with relatively little regard to species or plant part. The white-tailed deer feeds while moving, taking a succulent leaf here and a twig there, and is a good example of a gleaner.

It is the swathers that proved successful as domesticates and the swathers that have the greatest future potential. The American bison has orders of magnitude more potential for domestication than does the white-tailed deer. Even broader food niches were selected for in domestic stock, and niche differences between the sexes would be quickly obliterated. It is likely that swathers show much less difference in niche separation between the sexes than gleaners do in the wild state; broad food habits in similar vegetation require little discrimination, while specialized food habits in complex vegetation require considerable discrimination. Thus, domestic animals derived from swather wild stock would allow ready substitution of the sexes in population composition. Domestic cattle are swathers of grasses, sheep are swathers of grasses or gleaners of woody vegetation, and goats are gleaners of woody vegetation. The distinctions are broad, but useful. Cattle will eat some woody vegetation if grasses are sparse, and goats will eat grass if it is available, yet all three species have catholic food habits. Animal production will be greatest where the vegetation is most homogeneous in structure to be swathed; i.e., grasslands. The history of man has been characterized by attempts to convert woody vegetation to grasslands for grazing animals. Traditionally, fire was the tool, but recently mechanical and chemical means have been added.

These considerations lead to the following predictions about the biological efficiency of domestic animal management versus game cropping for meat production. In those areas where swathing is feasible (either grasslands or kinds of vegetation that can be converted to grasslands), domestic animals probably will produce a greater yield in weight than will the cropping of wild ungulates. In heterogeneous environments not readily cropped by swathing, sustained yield harvesting of the mixed native ungulates will probably yield a greater weight of meat than will ranching domestic stock. The economic trade-offs of costs of managing domestic animals versus the cost of harvesting and marketing game will vary by area, climate, whether meat is used for local subsistence or in a cash economy, distance to market, demand, and numerous other variables. Social attitudes of the producers will figure in the decision. Not all economic values of wildlife come from meat production. Ultimately, all factors—biological, economic, political, social, and aesthetic—must be weighed before deciding which method may be the most beneficial to man.

Integrating Social and Biological Factors

In the United States the enabling legislation of most states, establishing a commission with responsibility for management of wildlife, prescribes that wildlife will be managed for the maximum benefit of the people. The commissioners, usually appointed by the governor (although the practice varies from state to state), are responsible for setting policies that serve the interests of the people of the state. Departments of wildlife or game management were established to develop information to guide decisions by the commission, and to carry out the programs which resulted.

Two historical factors are fundamental to understanding the development of wildlife management in North America since the turn of the century. First, the field got its start when the nation was predominantly rural and agrarian, in both physical and philosophical terms. The attitude toward wildlife was largely utilitarian. To be sure, aesthetic and recreational aspects entered in, but the measure of success was game in the bag. Second, at both state and federal levels, the programs of management were financed by hunters. Thus, the constituency taxed was also that served, and the constituency was relatively homogeneous in its viewpoint. It is predictable that once wildlife populations had recovered from the low populations of the early 1900s (Allen, 1954), the thinking of wildlife biologists would tend to be toward maximizing the kill. This, in conjunction with the realization that the conservative kills maintained during the restoration of population numbers were much below what the population could support, led deer biologists to advocate liberalization of the kill. Furthermore, the growing signs of habitat destruction and winter mortality convinced biologists that action was vitally needed. It was recognized that population control could not be achieved by shooting bucks only, and the taking of antlerless deer was advocated. Carrying capacity became the new watchword.

Schisms developed, both within the game departments and in the hunting public. One group believed that restoration was complete (in fact, too complete) and that the program now required adequate exploitation of the population. The opposing group doubted that there were too many deer or, for that matter, that there could be too many deer, and feared that liberalized hunting would decimate the populations reestablished with such great effort. Lines were drawn which still exist. The situation has been further complicated by the growth, over the last decade, of a protectionist philosophy which is antihunting in viewpoint. Understandably enough, game commissions catered to what they perceived to be their public—the hunter, who conveniently also footed the bill. Protectionists objected to the emphasis on game species to the neglect of nongame wildlife, and this probably furthered their philosophical position that hunting was undesirable.

Coping with the biological questions is problem enough, but the controversy over wildlife management has now become emotional, as well. And since the basis for personal beliefs is deep and philosophical, it is not very amenable to education. For example, shooting bucks only was a major regulation during the recovery phase of deer populations from low levels at the end of the last century. The rationale was that in a polygynous species, males could be removed and females would still be saved for propagation. The selling of this idea involved more than the biological rationalization. It became a code of ethics of deer hunting. It was appropriate that hunters pursue only trophy males, the bigger, the better. The trophy, always a strong element in hunting satisfaction, was further emphasized, and the status of the hunter was based upon his ability to take trophy males. Shooting of females or young was taboo among legitimate sportsmen. Is it any wonder that a considerable number of hunters viewed the reinstitution of legal status for females and young with abhorrence? Or that any hunter who returned with a doe or a fawn in the bag would be regarded with contempt or ridicule? Or that such peer-group pressure, stemming from a deep-seated sense of ethics, should be a major barrier to communication?

There is a tendency among professional managers and biologists to assume that the resistance of the anti-antlerless hunter to sound biological information is traceable to stupidity. It is not. The point is, their train of logic starts from a different premise. Many opponents understand what the professional is saying, but they do not accept his evidence nor do they accept his conclusion. And, in fairness to the recalcitrant anti-antlerless hunting sportsman, it must be admitted that the evidence presented is usually subject to question in terms of statistical validity, and it is seldom so unambiguous as to disallow alternative interpretations.

While anti-antlerless hunting groups tend to be older, conservative, rural or north woods people (Hendee, 1969) the antihunting population tends to be young, liberal, and urban (Shaw, 1975). They are well-educated (many are amateur naturalists), politically astute, articulate, and outspoken. The antidoe shooters talk to each other; the antihunter groups talk to the uncommitted public. They perceive that since only about 10 percent of the adult population hunts, it is an activity that is quite vulnerable to the political process. Although the motivations of the antihunting groups vary considerably, a common (and probably fundamental) denominator is the objection to the killing of animals. Purists among them would object to causing the death of any animal and follow vegetarian habits. Others are of the humane group, objecting to any manner of death that causes pain and suffering. Sportsmen groups and their associated press and economic units emphasize the ideal sportsman and imply that he is typical. The antihunting groups emphasize the worst elements among hunters and imply that they are typical. Both exist, but neither is typical. Hunters, like any other more-or-less diverse population, include the extremes of ideal and disreputable with most members falling somewhere in the middle. Neither side seems particularly inclined to determine the actual distribution, so the rhetorical battle is likely to continue as an exchange of salvos from their respective philosophical promontories. Sportsmen emphasize that it is the total recreational experience including the aesthetic value of being close to nature that is the fundamental aspect of hunting. The antihunter counters that this can be obtained by nature study, hiking, photography, and a number of other pursuits that do not kill the animal. The sportsman replies that the taking of the animal (or at least its potential) is a necessary validation of the ritual. The antihunter retorts that killing is the fundamental motivation and, as such, is unbecoming a civilized society. The sportsman replies with arguments about the predatory nature of man, and that hunting is an instinct which, if thwarted, may result in a more serious disruption of society. Antihunters point out that other philosophers and scientists disagree with these viewpoints and that relatively few people hunt, yet most are no more antisocial than hunters. The hunter argues the utilitarian aspect of hunting, that the meat would otherwise be wasted. The antihunter replies that the meat is not needed; that alternative sources of meat are available at the market at less cost and effort; and that the waste of wild animals represents an egocentric human bias, since scavengers and decomposers need to live and the nutrients are more directly returned to the ecosystem without contributing to pollution problems. The sportsman argues that hunting is necessary to prevent populations from destroying their habitat and starving to death in the process. The antihunter replies that natural predator populations were suppressed because they competed with the hunter, and that they should be encouraged to resume their roles; further, that nature has marvelous recuperative powers, and periodic imbalances are no threat to the long-term balance if nature is only allowed to take its course. Other arguments abound, such as whether it is more humane to shoot an animal or let it be killed by a predator or starve to death. Ancillary issues, like gun control, become flanking movements. The mind-sets of the two groups appear irreconcilable.

It is hoped that by way of this rather lengthy presentation the results of this study can be placed in broader perspective. These results are biological results, whereas the major problems confronting the wildlife profession are philosophical and social. The biology may set extreme constraints that can be ignored only at considerable cost but the latitude is very great. There is no right answer, and management decisions must be made on arbitrary grounds according to relative values which a commission (or other decision-making body) perceives as serving the best interests of society. Furthermore, because the issues in the conflict stem from fundamental philosophical notions, it is unlikely that any compromise will constitute true conflict resolution. Compromise will be viewed by a given interest group as the best interim state that can be achieved, given their current power base. They may be temporarily satisfied with their gains or losses in the current skirmish, but efforts will continue in anticipation of future confrontations. The lot of commissioners and wildlife departments in this no-win game is not an enviable one.

Optimizing Benefits

The recognition that carrying capacity represents a range of values from I to K illustrates the latitude available to the decision maker. Deer populations can be reduced below I if regulated carefully, or if adequate escape refugia are present where the deer are invulnerable. Carrying capacity is not a single value, somehow sanctioned by nature, that can be looked to as the right way to manage a deer population.

One can manage for MSY and thereby maximize the number of deer in the bag. This will result in the greatest success rate in terms of the number of hunters taking home an animal, but given the other recreational and aesthetic values of the hunt, it may not maximize hunter satisfaction. And, of course, no kind of hunting will give satisfaction to the antihunter. Managing for MSY essentially eliminates chronic losses and minimizes losses to natural predators. It maximizes yield, as well as population stability, and to the wildlife manager it represents the maximal control and conformity to population models and, therefore, predictability is high. Analysis of kill data is fruitful since hunting mortality accounts for a large portion of the total mortality. Management for MSY has, however, several disadvantages. It reduces standing crop in the field (even at the peak i.e., prehunt, annual population) (fig. 8.2) below that level obtained nearer to K carrying capacity. This phenomenon has been at the root of much of the failure of communication between deer-management biologists and anti-antlerless deer hunters. The biologist uses the standard of a sustained, high level of harvest over many years as the measure of the success of his program. He emphasizes that recruitment increases to offset the higher hunting take under heavy-harvest programs.

The deer hunter both individually and collectively (at least within the peer group) uses a different standard. A high total kill (or percentage of hunter success) is a meaningless statistic to the individual if he failed to kill a deer. There is a very human tendency among hunters to regard the failure to kill a deer as due to few deer or bad luck, while success is attributed to skill. Failure is a source of discontent, but the converse is not necessarily true. Success, being attributed to skill, is not equated with success of the deer-management program. Thus, the hunter, even though personally successful, may be dissatisfied with the management program. One need not search long to discover that the single most important criterion of the hunter as to the satisfactory state of the deer herd is the number of deer seen in the field. If tracks and droppings are abundant and deer are seen regularly, the hunter is satisfied. If few deer are seen, even the successful hunter is likely to be disgruntled. The point was brought home most forcefully by a television program in which a sportsmen's group was displaying mounted trophy bucks taken by the winners in a competition. It is hard to imagine a group of hunters that should have been more satisfied than the winners. But in response to the interviewer's question of whether they regarded the season as successful, a majority of the succesful trophy hunters responded negatively. The reason given was that relatively few deer were seen while hunting. How often has one heard the lament, "I've hunted such-and-such an area for *X* number of years, and there are nowhere near as many deer as there used to be.''

The sad part of this impasse is that both sides are correct, but each by its own set of measures. Once again we encounter the I- versus K-carrying-capacity discrepancy, only in a different guise. Changes in the quantity and quality of deer habitat have occurred over the period of this conflict and have had obvious bearing upon the interpretation of events. But few, if any, deer biologists have recognized that management of a deer population for MSY inevitably results in lowering the standing crop in the field, even at peak, prehunt populations. Complete compensation of hunting removal by increased recruitment occurs only close to K carrying capacity. As one moves towards I carrying capacity, the peak popultion declines (fig. 8.2). This factor has several ramifications for the individual hunter. First, the number of deer in the field will be lower and the probability of seeing deer will be reduced. Second, the effort expended per deer killed will incease. Thus, while the percentage of hunters taking animals will increase, successful hunters, on the average, will have to work much harder for their deer. These circumstances are not likely to yield satisfaction to hunters with the anti-antlerless mind-set. Another disadvantage of management for MSY is that such low deer densities allow accelerated rates of succession. In a subclimax species such as the white-tailed deer, this is an important variable. Although the details from the George Reserve will be presented elsewhere, it is clearly the case (even though the rate is slow) for open fields being filled in by woody vegetation. The implications are obvious: maximization of yield of a subclimax ungulate is a short-term process if not combined with a program of habitat maintenance.

At the opposite extreme, one could completely eliminate legal human hunting. This option would satisfy the antihunting groups but obviously would be an extremely bitter event to most hunters. It seems reasonable to expect substantial illegal hunting in defiance of such a move, but for purposes of discussion, we will assume that the level of poaching remains unchanged. One can expect that the deer population would increase to K carrying capacity. The high density would result in maximum stabilization of successional rates. However, substantial population fluctuation could be expected (fig. 9.2) and the occurrence of catastrophic events would become more regular. Reestablishment of viable populations of the effective natural predators—the wolf and mountain lion—would introduce a stabilizing influence, but these animals are incompatible with current land-use practices in most of the white-tailed deer range, or most of settled North America for that matter. Therefore, it would appear to be an impractical solution. Without question, antifertility agents, live trapping and transplanting, and similar artificial measures would be tried, as witness recent absurdities with local deer populations close to the public eye. As a longtime observer of the white-tailed deer, I am confident of their ability to thwart or overwhelm the best efforts in that direction. Population fluctuations of large magnitude and unpredictable timing seem to be inevitable given total protection. Also, many sources of information on the status of the population would be unavailable (unless large-scale collections for research were allowed), although the need for such information would be greatly reduced.

Moderate harvests, such as are achieved by bucks-only or with modest numbers of antlerless animals, result in intermediate situations. Bucks-only hunting, particularly in situations where yearling males are excluded by legal definition, results in quite low removals. High standing crops in the field slow or stop succession and deer are readily observable, a plus for keeping the anti-antlerless deer hunter happy. Also, in areas where outdoor recreation and tourism are important, the probability of seeing deer as part of the outdoor experience is high. A good complement of natural predators and scavengers can be maintained on the relatively large numbers of deer reaching old age.

The disadvantages are related to the fact that such management programs result in a population near K carrying capacity. Natural fluctuations will occur and the probability of catastrophic events will be high. Data from the kill yields little information about the status of the population, and the proportion of the total mortality accounted for by the harvest is minor. Encouraging a healthy, large predator population where possible, or supplementing the bucks-only kill with a moderate, antlerless take can introduce greater stability in the population and reduce the probability and/or degree of impact of catastrophic events. The amount of stability introduced will depend upon the size of the removal. Of course, stability comes at the cost of lowered standing crop in the field and increased rates of succession. Again, the trade-offs are feasible throughout the range from I to K.

A rational program of deer (or any other game animal) management must start with the establishment of objectives. What segments of the public will benefit or what are the trade-offs between special interest groups? What dissatisfactions will be borne as part of the cost of carrying out the program? A commission may decide to manage for different interest groups in different areas. For example, hunting could be completely restricted in wilderness areas where natural predators and a complete ecological unit are present. Populations could be allowed to follow their dynamic natural equilibrium. Back-country areas, that serve as buffer zones between wilderness and man-altered ecosystems, could be used for bucks-only hunting. Areas of ready human access and traditionally good hunting areas could be managed for MSY or FRY. Areas of conflict between deer- and land-use patterns, such as farm or garden damage and high deer-car collisions, could be managed for SY below I carrying capacity.

Once objectives have been set, the general outline of management practices will have been established. Past experience and biological data in hand will suggest the initial action required (if any) to move in the direction of the desired objective. Still, it is important to establish, at the beginning, the criteria by which the success of the management program will be evaluated. Will these criteria be acceptable to the interest group or groups to which the program is directed? If not, can the interest group be convinced by information and education, or can alternative or compromise criteria be found? If the public to be served does not accept the criteria, failure will be preordained.

Finally, the program should be pursued, deliberately and incrementally. Communication on this scale is slow and imperfect. Too often in the past, too much was attempted in too short a time, and loss of public confidence resulted. Long periods of bucks-only hunting were followed by crises-infused heavy antlerless kills. Hunters were frightened by the magnitude of such removals, and feared for the future of the deer population. This was a dear price to pay, for long-term success is dependent upon public support. Crisis situations do occur that demand drastic action, but they are rare and, under continued management, should be extinct. As a general rule, new steps should not be taken until the public served by them can be carried along. There has been an overwhelming tendency among wildlife professionals to stress the need to convince, to educate the public. It never seems to occur to them that they represent the narrowest of interest groups. Managers want to manage, but if war is too important to be left up to the generals, then wildlife management is too important to be left to wildlife managers. Left alone, they tend to become deer farmers, trying to put the maximum amount of venison on car fenders.

Wildlife management is, and always will be, an art. It serves a broad and complex set of human needs and touches every element of the human psyche, from satisfying appetitive drives for food, to the most ethereal spiritual experiences. We may never plumb the depths of some of these motivations. It will suffice to know they exist in bewildering profusion. The art of wildlife management is to make possible the satisfaction of the needs of society within the biological constraints of the natural world. Science can locate the boundaries of the constraints; the manager must find the place within those constraints in which human needs are best met. His success will be dependent upon his ability to listen. His finger must seek the pulse of society, and not his own.

CHAPTER 17

Summing Up

The Ecosystem Hypothesis

At the outset of this book, a broad, conceptual hypothesis of ecosystem function was presented. It will be a long time before its usefulness can be assesed, since it cannot be stated in a manner that leads to direct verification. It is not an operational hypothesis, but rather a concept from which to derive operational hypotheses that are amenable to testing.

The results of this long-term study of white-tailed deer conform to the proposed ecosystem concept. Deer population responses showed the expected characteristics. The dynamic interaction between deer and vegetation seemed to function in something like the manner conceived. It is true that the deer is just one of a whole series of terrestrial herbivores on the George Reserve. Rodents and insects account for many more species and orders of magnitude of greater number of individuals, and they may not conform to the ideas presented in the concept. But in terrestrial ecosystems, K-selected species of plants and animals dominate the system in ways which control the system and set constraints, within which the *r*-selected organisms must function. Obviously this dominance is not complete. In disturbed areas, *r*-selected organisms do well and "outbreaks" of *r*-selected organisms (called pests, parasites, and diseases if they affect us) periodically raise havoc with K-selected dominants. But such events are of relatively short duration, and dominance in the terrestrial system usually reverts to K-selected organisms. The fact that it has never been deemed necessary to artificially control any of the other herbivores for the protection of the vegetation on the George Reserve is good evidence of the dominant status of the K-selected deer. Also, the impact of the existing predator, man, seemed to produce the expected deer-population responses. And the examination of evidence of the impact of natural predators (including aboriginal man) on deer suggests that a dynamic equilibrium existed between deer and predator within the framework for a linked vegetation, herbivore, and predator interrelationship.

The Deer Population Model

In this work a great deal of effort was made to exclude environmental variables and to have the only major variable be the size of the posthunt population. In the field situation, of course, it is impossible to control all variables, but the relatively stable climate of southern Michigan and the slow, background rate of vegetation change on the George Reserve resulted in relatively narrow ranges of uncontrolled environmental variables. Indeed, during the course of the study, deer density seemed to be the major variable influencing vegetation change. Similarly, the changes in the controlled variable were planned to occur in discrete, orderly steps. As it turned out, the alteration of population in discrete steps was not achieved, but the overall progressive reduction of the posthunt population was. The experiment certainly was not elegant, but given the difficulties, it seems reasonable to consider it a qualified success.

The outcome was a derivation of a series of equations that could be linked into a reasonably predictive model of population dynamics over a range of system states. In some sections of the analysis, assumptions had to be made, and the specific approach could be criticized. But the results, taken in total, show a remarkable consistency. The necessary internal consistency of population size, sex and age structure, reproductive rate, and longevity, and the overwhelming impact of resource limitations were such that the results could not have been much different or the discrepancies would have been glaring. Obviously the numerical values obtained are valid only for the George Reserve. The overall responses, however, would seem to be general to large K-selected organisms.

The message to the ecologist, I believe, is that these populations do not respond in a random fashion. A considerable amount of variation in a given parameter may occur internally, but such variation between parameters is not independent. The population serves as an integrator and achieves approximately the expected end with an almost infinite combination of values of given population parameters. Moreover, the results caution against simplistic concepts of population regulation. Under given circumstances, a given single factor may predominate; but under other conditions, it will switch to an entirely different factor. The dynamic tendency of the interactions will be to equalize the impact of resources and enemies in population regulation.

To the manager, it should be some comfort to know that most of the hypotheses about population dynamics of deer can be demonstrated in a field situation, and that the rationale for management is not entirely fictional. It should also be helpful to know specifically how some of the processes interact and why some of the traditional wisdom is inadequate. The route to a stable program of management is to stabilize those variables that are within control. Population models will help the managers, but they will seldom be answers in themselves. They will not stop the inclination of legislators to slash budgets or pressure groups to howl for scalps. But the numbers they generate are stark and confronting; they demand an evaluation. It is inevitable that the questioning of assumptions and probing of possibilities will lead to clarification of the problem and improvement in the kinds and amounts of data collected. The process of analysis is the immediate payoff of the model. And perhaps, occasionally, models will grow and evolve to the point where they yield more answers than questions.

Future Research

The second phase of the research—verification of the population model at high deer densities—is already underway. Because of time lags and expected greater variation, increments in this direction will require a longer time to complete than those going toward lower densities.

Assuming the completion of that phase, a major question involves niche separation between the sexes and its impact on the population consequences of shifts in sex ratio. The tentative conclusion reached here, that yield cannot be greatly increased by unbalancing sex ratio in favor of females, is one that is readily testable on the George Reserve. The advantages of shifting age ratios can also be tested.

A number of important questions raised by this study cannot be tested on the George Reserve because of small sample sizes. For example, a test of the validity of the age-structure-analysis approach to population parameter estimation and the subsequent derivation of a population model cannot be done on the George

Reserve. Similarly, more precise studies of the relationship of embryo rates to recruitment rates and population density cannot be done because of sample-size problems in determining embryo rates. And, obviously, the presence of a deer-proof fence rules out studies of immigration and emigration as population phenomena.

Nevertheless, it is hoped that the present work has sharpened these questions in a way that stimulates their testing in other, more appropriate situations. The George Reserve deer herd has given some valuable answers, but in the long run, the questions it has raised may be the more valuable contribution.

References

Allen, D. L. 1954. *Our wildlife legacy*. New York: Funk and Wagnalls.

Alexander, R. D. 1974. The evolution of social behavior. *Annual Rev. of Ecology and Systematics* 5:325–83.

Anderson, F. S. 1961. Effect of density on animal sex ratio. *Oikos* 12:1–16.

Armstrong, R. A. 1950. Fetal development of the northern white-tailed deer (*Odocoileus virginianus borealis* Miller). *Amer. Midland Nat.* 43:650–66.

Asdell, S. A. 1964. *Patterns of mammalian reproduction*. 2d ed. London: Constable.

Baker, R. H. 1968. Habitats and distribution. *Biology of* Peromyscus *(Rodentia)*, ed. J. A. King, pp. 98–126. Amer. Soc. Mammal. Spec. Pub. No. 2.

Barash, D. 1977. *Sociology and behavior*. New York: Elsevier.

Bartlett, I. H. 1949. White-tailed deer resources, United States and Canada. *Trans. N. Amer. Wild. Conf.* 14:543–52.

Bennett, C. L., Jr.; Ryel, L. A.; and Hawn, L. J. 1966. *A history of Michigan deer hunting*. Michigan Dept. Cons. Res. and Dev. Rep. No. 85.

Bromley, D. D. 1968. A comparative study of three methods of aging the white-tailed deer. Master's thesis, University of Michigan.

Bowyer, R. T. In prep. The evolution of sexual dimorphism in ungulates.

Brown, J. L. 1964. The evolution of diversity in avian territorial systems. *Wilson Bull.* 76:160–69.

———. 1969. Territorial behavior and population regulation in birds. *Wilson Bull.* 81:293–329.

Cantrell, I. J. 1943. *The ecology of the* Orthoptera *and* Dermoptera *of the George Reserve, Michigan*. Misc. Pub. Mus. Zool. No. 54. Ann Arbor: University of Michigan Press.

Caughley, G. 1966. Mortality patterns in mammals. *Ecology* 47:906–18.

———. 1970. Eruption of ungulate populations, with emphasis on Himalayan thar in New Zealand. *Ecology* 51:53–72.

———. 1974. Interpretation of age ratios. *J. Wildl. Manage.* 38:557–62.

———. 1974. Productivity, offtake, and rate of increase. *J. Wildl. Manage.* 38:566–67.

———. 1976. Wildlife management and the dynamics of ungulate populations. In *Applied biology,* vol. 1, ed. T. H. Coaker, pp. 183–246. London: Academic Press.

———, and L. C. Birch. 1971. Rate of increase. *J. Wildl. Manage.* 35:658–63.

Chase, W. W., and Jenkins, D. H. 1962. Productivity of the George Reserve deer herd. *Natl. Deer Disease Symp.* 1:78–88.

Cheatum, E. L. 1949. The use of corpora lutea for determining ovulation incidence and variation in the fertility of white-tailed deer. *Cornell Vet.* 39:282–91.

———, and Morton, G. H. 1942. On occurrence of pregnancy in white-tailed deer fawns. *J. Mammal.* 23:210–11.

———, and Severinghaus, C. W. 1950. Variations in fertility of white-tailed deer related to range conditions. *Trans. N. Amer. Wildl. Conf.* 15:170–90.

Christian, J. J. 1963. Endrocrine adaptative mechanisms and the physiological regulation of population growth. In *Physiological mammalogy,* vol. 1, eds. W. V. Mayer and R. G. VanGelder, pp. 189–353. New York: Academic Press.

———, and Davis, D. E. 1964. Endocrines, behavior and population. *Science* 146:1550–60.

Clements, F. E. 1916. *Plant succession: analysis of the development of vegetation.* Washington: Carnegie Institute.

Coblentz, B. E. 1974. Ecology, behavior, and range relationships of the feral goat. Ph.D. dissertation, University of Michigan.

Cole, L. C. 1954. The population consequences of life history phenomena. *Quart. Rev. Biol.* 29:103–37.

Collias, N. E. 1956. The analysis of socialization in sheep and goats. *Ecology* 37:228–38.

Crawford, H. S. 1957. A study of some factors in the fawning complex of the white-tailed deer. Master's thesis, University of Michigan.

Crisler, L. 1956. Observations of wolves hunting caribou. *J. Mammal.* 37:337–46.

Croon, G. W.; McCullough, D. R.; Olson, C. E., Jr.; and Queal, L. M. 1968. Infrared scanning techniques for big game censusing. *J. Wildl. Manage.* 32:751–59.

Dasmann, R. F. 1964*a*. *Wildlife biology.* New York: John Wiley and Sons.

———. 1964*b*. *African game ranching.* Oxford: Pergamon Press, Ltd.

Deevey, E. S., Jr. 1947. Live tables for natural populations of animals. *Quart. Rev. Biol.* 22:283–314.

Downing, R. L. 1965. An unusual sex ratio in white-tailed deer. *J. Wildl. Manage.* 29:884–85.

Earhart, C. M., and Johnson, N. K. 1970. Size dimorphism and food habits of North American owls. *Condor* 72:251–64.

Eberhardt, L. L. 1960. *Estimation of vital characteristics of Michigan deer herds*. Michigan Dept. Cons. Game Div. Rep. 2282.

———. 1969. Population analysis. In *Wildlife management techniques*, ed. R. H. Giles, Jr., pp. 457–95. Washington, D.C.: Wildlife Society.

Edwards, R. Y., and Fowle, C. D. 1955. The concept of carrying capacity. *Trans. N. Amer. Wildl. Conf.* 20:589–602.

Elder, W. H. 1965. Primeval deer hunting pressure revealed by remains from American Indian middens. *J. Wildl. Manage.* 29:366–70.

Erickson, J. A.; Anderson, A. E.; Medin, D. E.; and Bowden, D. C. 1970. Estimating ages of mule deer—an evaluation of technique accuracy. *J. Wildl. Manage.* 34:523–31.

Errington, P. L. 1934. Vulnerability of bob-white populations to predation. *Ecology* 15:110–27.

Estes, R. D. 1974. Social organization of African Bovidae. In *The behavior of ungulates and its relation to management,* eds. V. Geist and F. Walther, pp. 166–205. IUCN Publ. New Series No. 24, vol. 1.

Feist, J. D., and McCullough, D. R. 1975. Reproduction in feral horses. *J. Reprod. Fert., Suppl.* 23:23–28.

———. 1976. Behavior patterns and communication in feral horses. *Z. Tierpsychol.* 41:337–71.

Fisher, R. A. 1958. *The genetical theory of natural selection.* 2nd rev. ed. New York: Dover.

Flook, D. R. 1970. *A study of sex differential in the survival of wapiti.* Canadian Wildl. Ser. Rep. Series No. 11.

Franklin, W. L. 1974. The social behavior of vicuña. In *The behavior of ungulates and its relation to management,* eds. V. Geist and F. Walther, pp. 447–87. IUCN Publ. New Series No. 24, vol. 1.

Freeman, D. C.; Klikoff, L. G.; and Harper, K. T. 1976. Differential resource utilization by the sexes of dioecious plants. *Science* 193:597–99.

Gadgil, M., and Bossert, W. H. 1970. Life history consequences of natural selection. *Amer. Nat.* 104:1–24.

Geist, V. 1971. *Mountain sheep: a study in behavior and evolution.* Chicago: University of Chicago Press.

———, and Petocz, R. G. 1977. Bighorn sheep in winter: do rams maximize reproductive fitness by spatial and habitat segregation from ewes? *Canadian J. Zool.* 55:1802–10.

Gilbert, F. F., and Stolt, S. L. 1970. Variability in aging Maine white-tailed deer by tooth-wear characteristics. *J. Wildl. Manage.* 34: 532–37.

Gross, J. E. 1969. Optimum yield in deer and elk populations. *Trans. N. Amer. Wildl. and Nat. Resour. Conf.* 34:372–85.

———. 1972. Criteria for big game planning: performance measures vs. intuition. *Trans. N. Amer. Wildl. and Nat. Resour. Conf.* 37:246–59.

Guilday, J. E. 1962. The deer of 350 years ago. *Pennsylvania Game News* 33:10–13.

———; Parmalee, P. W.; and Tanner, D. P. 1962. Aboriginal butchering techniques at the Eschelman Site (36 La 12) Lancaster County, Pennsylvania. *Archaeologist Bull.* 32:59–83.

Halls, L. K., and Crawford, H. S., Jr. 1960. Deer-forest habitat relationships in north Arkansas. *J. Wildl. Manage.* 24:387–95.

Hamilton, W. D. 1964. The genetical theory of social behavior. I and II. *J. Theor. Biol.* 7:1–52.

———. 1967. Extraordinary sex ratios. *Science* 156:477–88.

Haugen, A. O. 1975. Reproductive performance of white-tailed deer in Iowa. *J. Mammal.* 56:151–59.

Hawkins, R. E., and Klimstra, W. D. 1970. A preliminary study of the social organization of the white-tailed deer. *J. Wildl. Manage.* 34:407–19.

Hawn, L. J. 1974. *The 1974 deer seasons.* Michigan Dept. Nat. Resour., Surveys and Statistics Serv. Rep. No. 16.

Hendee, J. C. 1969. Appreciative versus consumptive uses of wildlife refuges: studies of who gets what and trends in use. *Trans. N. Amer. Wildl. and Nat. Resour. Conf.* 34:252–64.

Hesselton, W. T.; Severinghaus, C. W.; and Tanck, J. E. 1965. Population dynamics of deer at the Seneca Army Depot. *New York Fish and Game J.* 12:17–30.

———, and Jackson, L. W. 1974. Reproductive rates of white-tailed deer in New York State. *New York Fish and Game J.* 21:135–52.

Hirth, D. H. 1977. *Social behavior of white-tailed deer in relation to habitat.* Wildl. Monogr. No. 53.

———, and McCullough, D. R. 1977. Alarm signals in ungulates with special reference to white-tailed deer. *Amer. Nat.* 111:31–42.

Hickie, P. 1937. Four deer produce 160 in six seasons. *Mich. Cons.* 7:6–7, 11.

Hornocker, M. G. 1970. *An analysis of mountain lion predation upon mule deer and elk in the Idaho Primitive Area.* Wildl. Monogr. 21.

Hoskinson, R. L., and Mech, L. D. 1976. White-tailed deer migration and its role in wolf predation. *J. Wildl. Manage.* 40:429–41.

Huffacker, C. B. 1958. Experimental studies on predation: dispersion factors and predator-prey oscillations. *Hilgardia* 27:343–83.

Hunter, G. N. 1957. The techniques used in Colorado to obtain hunter distribution. *Trans. N. Amer. Wildl. Conf.* 22:584–93.

Ismond, M. D. 1952. A deer behavior study with particular reference to the pre-fawning period on the George Reserve, Michigan. Master's thesis, University of Michigan.

Jarman, P. J. 1974. The social organization of antelope in relation to their ecology. *Behaviour* 58:215–67.

Jenkins, D. H. 1964. The productivity and management of deer on the Edwin S. George Reserve, Michigan. Ph.D. dissertation, University of Michigan.

———. 1970. Harvest regulations and population control for midwestern deer. In *White-tailed deer in the Midwest,* pp. 23–27. USDA For. Serv. Res. Paper NC-39.

Jordan, P. A.; Shelton, P. C.; and Allen, D. L. 1967. Numbers, turnover and social structure of the Isle Royale wolf population. *Amer. Zool.* 7:233–52.

Kabat, C.; Collias, N. E.; and Guettinger, R. C. 1953. *Some winter habits of white-tailed deer and the development of census methods in the Flag Yard of northern Wisconsin.* Wisc. Cons. Dept. Tech. Wildl. Bull. 7.

Kammermeyer, K. K., and Marchinton, R. L. 1967. Notes on dispersal of male white-tailed deer. *J. Mamm.* 57:776–78.

Kaufmann, J. H. 1962. Ecology and social behavior of the coati, (*Nasua narica*) on Barro Colorado Island, Panama. *Univ. Calif. Pub. Zool.* 60:95–222.

Kelker, G. H. 1947. Computing the rate of increase for deer. *J. Wildl. Manage.* 11:177–83.

Kitchen, D. W. 1974. *Social behavior and ecology of the pronghorn.* Wildl. Monogr. 38.

Kleiber, M. 1961. *The fire of life.* New York: Wiley.

Klein, D. R. 1968. The introduction, increase, and crash of reindeer on St. Mathew Island. *J. Wildl. Manage.* 32:350–67.

Klingel, H. 1969. Reproduction in the plains zebra, *Equus burchelli boehmi:* behavior and ecological factors. *J. Reprod. Fert., Suppl.* 6:339–45.

Koford, C. 1957. The vicuña and the puna. *Ecological Monogr.* 27:153–219.

Kolenosky, G. G. 1972. Wolf predation on wintering deer in east-central Ontario. *J. Wildl. Manage.* 36:357–69.

Kolman, W. A. 1960. The mechanism of natural selection for the sex ratio. *Amer. Nat.* 94:373–77.

Krebs, C. J. 1972. *Ecology.* New York: Harper and Row.

Langenau, E. E., Jr. 1973. *An experimental analysis of aggression in penned white-tailed buck fawns during winter.* Michigan Dept. Nat. Resour., Wildl. Div. Rep. No. 2703.

———, and Lerg, J. M. 1976. The effects of winter nutritional stress on maternal and neonatal behavior in penned white-tailed deer. *Applied Anim. Ethology* 2:207–23.

Leigh, E. G., Jr. 1970. Sex ratio and differential mortality between the sexes. *Amer. Nat.* 104:205–10.

Leopold, A. 1933. *Game management.* New York: Charles Scribner's Sons.

———. 1943. Deer irruptions. *Trans. Wisc. Acad. Sci., Arts, and Letters* 35:351–66.

———; Sowls, L. K.; and Spencer, D. L. 1947. A survey of overpopulated deer ranges in the United States. *J. Wildl. Manage.* 11:162–77.

Leopold, A. S. 1950. Deer in relation to plant succession. *Trans. N. Amer. Nat. Resour. Conf.* 15:571–80.

Levins, R. 1968. *Evolution in changing environments: some theoretical explorations*. Princeton, N.J.: Princeton University Press.

Lidicker, W. Z., Jr. 1962. Emigration as a possible mechanism permitting the regulation of population density below carrying capacity. *Amer. Nat.* 96:29–33.

Lockard, G. R. 1972. Further studies of dental annuli for aging white-tailed deer. *J. Wildl. Manage.* 36:46–55.

Longhurst, W. H. 1957. The effectiveness of hunting in controlling big game populations in North America. *Trans. N. Amer. Wildl. Conf.* 22:544–69.

Lowe, V. P. W. 1969. Population dynamics of the red deer (*Cervus elaphus* L.) on Rhum. *J. Anim. Ecology* 38:425–57.

MacArthur, R. H. 1972. *Geographical ecology: patterns in the distribution of species*. New York: Harper and Row.

———, and Wilson, E. O. 1967. *The theory of island biogeography*. Monogr. in Population Biol., Princeton University Press.

Maguire, H. F., and Severinghaus, C. W. 1954. Wariness as an influence on age composition of white-tailed deer killed by hunters. *New York Fish and Game J.* 1:98–109.

Maiorana, V. C. 1976. Reproductive value, prudent predators, and group selection. *Amer. Nat.* 110:486–89.

Mansell, W. D. 1971. Accessory corpora lutea in ovaries of white-tailed deer. *J. Wildl. Manage.* 35:369–457.

Margalef, R. 1963. On certain unifying principles in ecology. *Amer. Nat.* 97:357–74.

May, R. M. 1976. Models for single populations. In *Theoretical ecology: principles and applications,* ed. R. M. May, pp. 4–25. Philadelphia: W. B. Saunders.

Maynard-Smith, J. 1968. *Mathematical ideas in biology*. New York: Cambridge University Press.

———, and Slatkin, M. 1973. The stability of predator-prey systems. *Ecology* 54:384–91.

McCullough, D. R. 1969. The tule elk: its history, behavior and ecology. *Univ. Calif. Pub. Zool.* 88:1–209.

———. 1970. Secondary production in birds and mammals. In *Analysis of temperate forest ecosystems,* ed. D. E. Reichle, pp. 105–30. Berlin: Springer-Verlag.

———. 1978. Essential data required on population structure and dynamics in field studies of threatened herbivores. In *Threatened Deer,* pp. 302–17. Morges: IUCN.

———; Olson, C. E., Jr.; and Queal, L. M. 1969. Progress in large animal census by thermal mapping. In *Remote sensing in ecology,* ed. P. L. Johnson, p. 138–47. Athens, Ga.: University of Georgia Press.

McDowell, R. D. 1959. *Relationship of maternal age to parental sex ratios in white-tailed deer.* Proc. N.E. Sec. Wildl. Soc.

McGinnis, B. S., and Downing, R. L. 1977. Factors affecting the peak of white-tailed deer fawning in Virginia. *J. Wildl. Manage.* 41:715–19.

———, and Reeves, J. H., Jr. 1958. Deer jaws from excavated Indian sites let us compare Indian-killed deer with modern deer. *Virginia Wildl.* 19:8–9.

McIntosh, R. P. 1963. Ecosystems, evolution and relational patterns of living organisms. *Amer. Sci.* 51:246–67.

McNab, B. K. 1963. Bioenergetics and the determination of home range size. *Amer. Nat.* 97:133–40.

McNeil, R. J. 1962. *Population dynamics and economic impact of deer in southern Michigan.* Mich. Dept. Cons. Game Div. Rep. 2395.

Mech, L. D. 1966. *The wolves of Isle Royale.* Fauna of the National Parks No. 7.

———. 1970. *The wolf: the ecology and behavior of an endangered species.* Garden City, N.Y.: Natural History Press.

———. 1975*a*. Population trend and winter deer consumption in a Minnesota wolf pack. In *Proc. 1975 predator symposium,* eds. R. L. Phillips and C. Jonkel, pp. 55–83. Montana For. and Cons. Exp. Stat., Univ. Montana, Missoula.

———. 1975*b*. Disproportionate sex ratios of wolf pups. *J. Wildl. Manage.* 39:737–40.

———. 1977*a*. Wolf-pack buffer zones as prey reservoirs. *Science* 198–320–21.

———. 1977*b*. Productivity, mortality, and population trend of wolves in northeastern Minnesota. *J. Mammal.* 58:559–74.

———, and Frenzel, L. D., Jr. 1971. An analysis of the age, sex and condition of deer killed by wolves in northeastern Minnesota. In *Ecological studies of the timber wolf in northeastern Minnesota,* eds. L. D. Mech and L. D. Frenzel, Jr., pp. 35–51. U.S. For. Serv. Res. Paper NC–52.

———; Frenzel, L. D.; and Karns, P. D. 1971. The effect of snow conditions on the vulnerability of white-tailed deer to wolf predation. In *Ecological studies of the timber wolf in northeastern Minnesota,* eds. L. D. Mech and L. D. Frenzel, Jr., pp. 51–59. U.S. For. Serv. Res. Paper NC–52.

———, and Karns, P. D. 1977. *Role of the wolf in a deer decline in the Superior National Forest.* U.S. For. Serv. Res. Paper NC–148.

Menzel, K. E. 1958. Seasonal and annual changes in weights of George Reserve deer. Master's thesis, University of Michigan.

Mertz, D. B. 1970. Notes on methods used in life-history studies. In *Readings in ecology and ecological genetics,* eds. J. H. Connell; D. B. Mertz; and W. W. Murdock, pp. 4–17. New York: Harper and Row.

———, and Wade, M. J. 1976. The prudent prey and the prudent predator. *Amer. Nat.* 110:489–96.

Miller, R. 1976. Models, metaphysics, and long-lived species. *Bull. Ecological Soc. Amer.* 57:2–6.

Mitchell, B. 1967. Growth layers in dental cementum for determining the age of red deer (*Cervus elaphus* L.). *J. Anim. Ecology* 36:279–93.

Moen, A. N. 1973. *Wildlife ecology*. San Francisco: W. H. Freeman.

Moran, P. A. P. 1950. Some remarks on animal population dynamics. *Biometrics* 6:250–58.

Morton, G. H., and Cheatum, E. L. 1946. Regional differences in breeding potential of white-tailed deer in New York. *J. Wildl. Manage.* 10:242–48.

Murie, A. 1944. *The wolves of Mount McKinley*. Fauna of the National Parks, No. 5.

Murphy, D. A. 1966. Effects of various opening days on deer harvest and hunting pressure. *Proc. S. E. Assoc. Game and Fish Comm.* 19:141–46.

———. 1969. Hunting methods, limits, and regulations. In *White-tailed deer in the southern forest habitat*, pp. 54–58. Proc. of a Symposium at Nacogdoches, Texas, March 25–26, 1969.

Myers, J. H., and Krebs, C. J. 1971. Sex ratios in open and enclosed vole populations: demographic implications. *Amer. Nat.* 105:325–44.

Neff, D. J. 1968. The pellet-group count technique for big game trend, census, and distribution: a review. *J. Wildl. Manage.* 32:597–614.

Newhouse, S. J. 1973. Effects of weather on behavior of white-tailed deer of the George Reserve, Michigan. Master's thesis, University of Michigan.

Odum, E. P. 1969. The strategy of ecosystem development. *Science* 164:262–70.

———. 1971. *Fundamentals of ecology*. 3rd ed. Philadelphia: W. B. Saunders.

O'Pezio, J. P. 1978. Mortality among white-tailed deer fawns on the Seneca Army Depot. *New York Fish and Game J.* 25:1–15.

O'Roke, E. C., and Hammerstrom, F. N., Jr. 1948. Productivity and yield of the George Reserve deer herd. *J. Wildl. Manage.* 12:78–86.

Ozoga, J. J. 1972. Aggressive behavior of white-tailed deer at winter cuttings. *J. Wildl. Manage.* 36:861–68.

———. 1969. Some longevity records for female white-tailed deer in northern Michigan. *J. Wildl. Manage.* 33:1027–28.

Pearl, R. 1930. *Introduction to medical biometry and statistics*. 2d ed. Philadelphia: W. B. Saunders.

Pianka, E. R. 1970. On r- and K-selection. *Amer. Nat.* 104:592–97.

———. 1972. r- and K-selection or b and d selection? *Amer. Nat.* 106:581–88.

Pimental, D. 1968. Population regulation and genetic feedback. *Science* 159:1432–37.

Pimlott, D. H.; Shannon, J. A.; and Kolenosky, G. B. 1969. *The ecology of the timber wolf in Algonquin Provincial Park*. Ontario Dept. Lands and For.

Queal, L. M. 1962. Behavior of white-tailed deer and factors affecting social organization of the species. Master's thesis, University of Michigan.

Quick, H. F. 1960. Animal population analysis. In *Manual of game investi-*

gational techniques, ed. H. S. Mosby, Washington, D.C.: The Wildlife Society.

Rabb, G. B.; Wooply, J. H.; and Ginsburg, B. E. 1967. Social relationships in a group of captive wolves. *Amer. Zool.* 7:305–11.

Ransom, B. A. 1966. Determining age of white-tailed deer from layers in cementum of molars. *J. Wildl. Manage.* 30:197–99.

Ricker, W. E. 1954. Stock and recruitment. *J. Fish. Res. Board Canada* 11:559–623.

———. 1975. *Computation and interpretation of biological statistics of fish populations.* Fish. Res. Board Canada Bull. 191.

Robinette, W. L. 1966. Mule deer home range and dispersal in Utah. *J. Wildl. Manage.* 30:335–49.

———; Gashwiler, J. S.; Jones, D. A.; and Crane, H. S. 1955. Fertility of mule deer in Utah. *J. Wildl. Manage.* 19:115–36.

———; Gashwiler, J. S.; Low, J. B.; and Jones, D. A. 1957. Differential mortality by sex and age among mule deer. *J. Wildl. Manage.* 21:1–16.

———; Gashwiler, J. S.; and Morris, O. W. 1959. Food habits of the cougar in Utah and Nevada. *J. Wildl. Manage.* 23:261–73.

Robinson, W. L. 1962. Social dominance and physical condition among penned white-tailed deer. *J. Wildl. Manage.* 13:195–216.

Roller, N. E. G. 1974. Airphoto mapping of ecosystem development on the Edwin S. George Reserve. Master's thesis, University of Michigan.

Roseberry, J. L., and Klimstra, W. D. 1974. Differential vulnerability during a controlled deer harvest. *J. Wildl. Manage.* 38:499–507.

Rosenzweig, M. L. and MacArthur, R. H. 1963. Graphical representation and stability conditions of predator-prey interactions. *Amer. Nat.* 97:209–23.

Ryel, L. A. 1971. *Evaluation of pellet group surveys for estimating deer populations in Michigan.* Mich. Dept. Nat. Resour., Res. and Dev. Rep. No. 250.

———. 1974. *The 1970 deer seasons.* Mich. Dept. Nat. Resour., Res. and Dev. Rep. No. 251A.

———; Fay, L. D.; and VanEtten, R. C. 1961. Validity of age determination in Michigan deer. *Mich. Acad. Sci. Arts and Letters.* 47:289–316.

Sadleir, R. M. F. S. 1969. *The ecology and reproduction in wild and dometic mammals.* London: Methuen and Co., Ltd.

Salzen, E. A. 1962. Imprinting and fear. *Symp. Zool. Soc. London* 8:199–217.

Scott, R. F. 1954. Population growth and game management. *Trans. N. Amer. Wildl. Conf.* 19:480–503.

Schladweiler, P. 1974. Response (to Caughley, 1974). *J. Wildl. Manage.* 38:567.

Seal, U. S.; Mech, L. D.; and Van Ballenberghe, V. 1975. Blood analysis of wolf pups and their ecological metabolic interpretation. *J. Mammal.* 56:64–75.

Seber, G. A. 1973. *Estimation of animal abundance*. New York: Hafner.
Selander, R. K. 1966. Sexual dimorphism and differential niche utilization in birds. *Condor* 68:113–51.
Severinghaus, C. W. 1949. Tooth development and wear as criteria of age in white-tailed deer. J. Wildl. Manage. 13:195–216.
———, and Cheatum, E. L. 1956. Life and times of the white-tailed deer. In *The deer of North America,* ed. W. P. Taylor, pp. 57–186. Harrisburg: Stackpole Co.
Shaw, W. W. 1975. *Attitudes toward hunting*. Mich. Dept. Nat. Resour. Wildl. Div. Rep. No. 2740.
Short, H. L. 1963. Rumen fermentations and energy relationships in white-tailed deer. *J. Wildl. Manage*. 27:184–95.
Slobodkin, L. B. 1968. How to be a predator. *Amer. Zool.* 8:43–51.
———. 1974. Prudent predation does not require group selection. *Amer. Nat*. 108:665–78.
Smith, B. D. 1975. *Middle Mississippi exploitation of animal populations*. Univ. of Mich., Mus. Anthro. Papers, No. 57.
Smith, F. E. 1975. Ecosystems and evolution. *Bull Ecological Soc. of Amer*. 56:2–6.
Smythe, N. 1970. The adaptive value of the social organization of the coati (*Nasua narica*). *J. Mammal.* 60:818–20.
Sokal, R. R., and Rohlf, F. J. 1969. *Biometry*. San Francisco: W. H. Freeman.
Southworth, T. R. E.; May, R. M.; Hassell, M. P.; and Conway, G. R. 1974. Ecological strategies and population parameters. *Amer. Nat.* 108:791–804.
Stenlund, M. H. 1955. *A field study of the timber wolf* (Canis lupus) *on the Superior National Forest, Minnesota*. Minn. Dept. Cons. Tech. Bull. No. 4.
Storer, R. W. 1966. Sexual dimorphism and food habits in three North American accipiters. *Auk* 83:423–36.
Stubbs, M. 1977. Density dependence in the life-cycles of animals and its importance in K- and *r*-selected strategies. *J. Anim. Ecology* 46:677–88.
Taber, R. D. 1953. The secondary sex ratio in *Odocoileus*. *J. Wildl. Manage*. 17:95–96.
———, and Dasmann, R. F. 1954. A sex difference in mortality in young Columbian black-tailed deer. *J. Wildl. Manage*. 18:309–14.
———, and Dasmann, R. F. 1958. *The black-tailed deer of the Chaparral.* Calif. Dept. Fish and Game, Game Bull. No. 8.
Tanner, J. T. 1966. Effects of population density on growth rates of animal populations. *Ecology* 47:734–40.
Taylor, W. P., ed. 1956. *The deer of North America*. Harrisburg: Stackpole Co.
Teer, J. G.; Thomas, J. W.; and Walker, E. A. 1965. *Ecology and management of white-tailed deer in the Llano Basin of Texas*. Wildl. Monogr. 15.

Terman, C. R., and Sassman, J. F. 1967. Sex ratio in deer mouse populations. *J. Mammal.* 48:589–97.
Tester, J. R., and Heezen, K. L. 1965. Deer response to a drive census determined by radio tracking. *BioScience* 15:100–104.
Thomas, D. C., and Bandy, P. J. 1975. Accuracy of dental-wear age estimates of black-tailed deer. *J. Wildl. Manage.* 39:674–78.
Trivers, R. L. 1971. The evolution of reciprocal altruism. *Quart. Rev. of Biol.* 46:35–37.
———. 1972. Parental investment and sexual selection. In *Sexual selection and the descent of man,* ed. B. Campbell, pp. 136–79. Chicago: Aldine.
———. 1974. Parent-offspring conflict. *Amer. Zool.* 14:249–64.
———, and Willard, D. E. 1973. Natural selection of parental ability to vary the sex ratio of offspring. *Science* 179:90–92.
Van Ballenberghe, V.; Erickson, A. W.; and Bymun, D. 1975. *Ecology of the timber wolf in northeastern Minnesota.* Wildl. Monogr. No. 43.
———, and Mech, L. D. 1975. Weights, growth, and survival of timber wolf pups in Minnesota. *J. Mammal.* 56:44–63.
Verme, L. J. 1962. Mortality of white-tailed deer fawns in relation to nutrition. *Proc. Natl. Deer Disease Symp.* 1:15–32.
———. 1963. Effect of nutrition on growth of white-tailed deer fawns. *Trans. N. Amer. Wildl. Conf.* 28:431–43.
———. 1965. Reproduction studies on penned white-tailed deer. *J. Wildl. Manage.* 29:74–79.
Verner, J. 1965. Selection for sex ratio. *Amer. Nat.* 99:419–21.
Voigt, D. R.; Kolenosky, G. B.; and Pimlott, D. H. 1976. Changes in summer foods of wolves in central Ontario. *J. Wildl. Manage.* 40:663–68.
Walters, C. J., and Bandy, P. J. 1972. Periodic harvest as a method of increasing big game yields. *J. Wildl. Manage.* 36:128–34.
Wagner, F. H. 1969. Ecosystem concepts in fish and game management. In *The ecosystem concept in natural resource management,* ed. G. W. Van Dyne, pp. 259–307. New York: Academic Press.
Weins, J. A. 1966. On group selection and Wynne-Edwards' hypothesis. *Amer. Scientist* 54:273–87.
Wilbur, H. M.; Tinkle, D. W.; and Collins, J. P. 1974. Environmental certainty, trophic level and resource availability in life history evolution. *Amer. Nat.* 108:805–17.
Williams, G. C. 1957. Pleiotropy, natural selection, and the evolution of senescence. *Evolution* 11:398–411.
———. 1966. *Adaptation and natural selection.* Princeton: Princeton University Press.
Wilson, E. O. 1975. *Sociobiology: the new synthesis.* Cambridge, Mass.: Harvard University Press, Belknap Press.
———, and Bossert, W. H. 1971. *A primer of population biology.* Stamford, Conn.: Sinauer.
Wolfe, M. L., and Allen, D. L. 1973. Continued studies of the status,

socialization, and relationships of Isle Royale wolves, 1967 to 1970. *J. Mammal.* 54:611–35.

Wynne-Edwards, V. C. 1962. *Animal dispersion in relation to social behavior.* New York: Hafner.

Youatt, W. G.; Fay, L. D.; and Harte, H. D. 1975. *1975 spring deer survey doe productivity and femur fat analysis April 1–June 1.* Mich. Dept. Nat. Resour. Rep. No. 2743.

———; Verme, L. J.; and Ullrey, D. E. 1965. Composition of milk and blood in nursing white-tailed does and blood composition of their fawns. *J. Wildl. Manage.* 29:79–84.

Zimen, E. 1976. On the regulation of pack size in wolves. *Z. Tierpsychol.* 40:300–41.

Index